5

A Perfect Stitch

The Extraordinary Singer 201 'Jubilee' Sewing Machine

By
Alex Askaroff

A Perfect Stitch

The Extraordinary
Singer 201
'Jubilee'
Sewing Machine

By
Alex Askaroff

On The Road Series

There are seven books in Alex Askaroff's **On The Road Series**. They cover his working life around Sussex encompassing a world of stories from the ages.

Book One: Patches of Heaven

Book Two: Skylark Country

Book Three: High Streets & Hedgerows

Book Four: Tales From The Coast

Book Five: Have I Got A Story For You

Book Six: Glory Days

Book Seven: Off The Beaten Track

Amazon print for worldwide release 2022
Paperback ISBN: 9798789426159

SINGER FACTORY, CLYDEBANK, SCOTLAND

Dedication

I have always wanted grandchildren. By December of 2021 amazingly I had three little darlings to tease. I dedicate this small book to my treasures, Evelyn, Alena and Rose. May they have wonderful lives.

Foreword

Alex Askaroff grew up in the sewing industry and has spent a lifetime studying and writing about his craft. He is a world renowned expert on pioneering sewing machine inventors, creating Sewalot.com which has become the premier website for antique sewing machines with millions of visitors. For decades he has assisted and consulted with novels, programs and films, from the BBC The Repair Shop to The Great British Sewing Bee, The Singer Story, Made in Clydebank, to How The Victorians Built Britain with Michael Buerk and more. His expertise in the sewing field has helped countless enthusiasts and collectors, giving interviews and writing numerous articles for magazines and publications worldwide. Alex Askaroff has had Nine No1 New Releases on Amazon.

NOTE:

Throughout this book Alex will keep coming back to the price of the 201. Why? Because it is the Singer 201s greatest enigma. How could a basic machine, that did the same as so many cheaper models, still remain as Singers flagship model for nearly 30 years! Between these pages you will discover why.

INDEX

Introduction

My relationship with the Singer 201 started early. Down in our cellar at 7 Ashburnham Gardens in Eastbourne was a 201k-1 treadle. It was the machine that my mum started her business with. A skilled Viennese seamstress by trade she had been taught by her mum, my British granny, who had run away from home and married an Austrian. She was an haute couture, high fashion dressmaker in Vienna after WW2 with her own shop. I still have granny's first machine, a Singer 28k in my loft.

After some disagreements with a Russian officer over the price of his mistress's evening gown, granny decided it might be best to grab the family and head for home. Granny went to live in Brighton and my parents built up a multi-million pound business in Eastbourne. However, mum could never part with her beloved 201. So, although surplus to requirements, into the cellar it went.

Eventually I completely restored the machine. When I think back now it was probably the first full restoration I completed. When I became Production Director at the family firm, although we ran on the latest industrial sewing machines, the Singer 201 proudly sat on display in the waiting room, welcoming all to our manufacturing powerhouse. So you see, I was well aware from a child just how special the 201 was.

This will probably be my last academic work and as usual I will try and fill the pages with interesting snippets and facts. The problem is now that I am the wrong side of 60 and I find that more stuff is falling out of my brains than going in! If you ever want to see how I manage to write these books I explained it all in my book, The History Of Frister & Rossmann Sewing Machines. It's a bit of an eye opener.

I'll tell you something, spending a lifetime in the sewing industry has been a joy. When I was a kid manufacturing in Britain was still booming and the factories were a sight to behold. There were rows of sewing girls hammering away on industrial machines in short staccato machine gun bursts, cutters cutting, packers packing. The excitement, noise and action of a busy factory is a sight.

The sewing world gave me an insight into a few special sewing machines that the machinists loved. There was the trendsetting domestic Bernina 125. The Brother B755 industrial that stitched just about anything. With its self-oiling sump it could sew continuously on 24hr shift work seven days a week and love it. Some were set up to go so fast that they needed jet air coolers to stop the needles overheating. Then there was the amazing walking foot needle-feed Seiko LSW8BL, loved by everyone from tarpaulin makers to upholsterers. It could sew through plywood like paper. Then of course there was the tiny but beautiful, Kylie Minogue of sewing machines, the Singer Featherweight that some quilters (even today) dream about.

Out of all the countless thousands of machines, makes and models there was one that stood head and shoulders above the rest. Not because it was more powerful, not because it was faster, not because it looked stunning (like some of the 1950s Necchi machines) but because it was as close to perfection as was possible for a sewing machine. Even people who did not understand why, loved it.

That machine was the incomparable Singer 201. Even today, with all the wonderful technology available for basic stitching, that machine still shines. I've often use to laugh when describing the 201 and say that if you've never had an emotional moment with a sewing machine, then you have never sewn on a Singer 201.

Let me give you an example. Once upon a time there was a little specialist book binders along Seaside in Eastbourne, down a little road by the Leaf Hall built by William Leaf. They used industrial Singer 31k machines to sew in the pages of some of their books to the bindings. There was a row of six of these heavy, noisy, lumpy machines, all pounding away. When it was busy you couldn't hear yourself think!

One day, when they were rushing to complete an order, one of the 31's broke down. I was called in and I had to take the machine away. The clever machinist had managed to sew a pair of snips under a book and bent the needle-bar! Well, the only machine I had in my car was an old Singer 201k that a customer had given me as she could not lift it anymore (the early pig iron ones weigh around 36 lbs). I offered the company the machine just as a stand in until I could bring their big industrial back.

I left the machine with them thinking it would probably not do the job. However, a week later when I returned, I was amazed to find that the machine sewed perfectly without a single complaint. About a year went by and I was called back to the unit to service their machines. As I stood outside the works there was silence. Not a good thing for a manufacturing business. When I went inside I was confronted by a row of six Singer 201k's all sewing away in quiet harmony. The company had switched all their industrials to the fabulous 201 and all the girls had smiles on their faces.

So that's the 201, a quiet unassuming little piece of perfection. In this book I want to give you some of the history of the machine and some of the legends too.

It has been my long passion to encourage other enthusiasts around the world. All I know about the 201 is between these pages. Please forgive any mistakes, like I say I'm an old dog now and I do forget. It might not be everything there is to know about the 201, but it's a good start. Sit back and enjoy it with the same enthusiasm that it was written.

Chapter 1
The Jubilee Sewing Machine

NEW SINGER 201K

::

The Jubilee Sewing Machine

The Singer 201 was also known for a period as The Jubilee Sewing Machine as it was officially launched in the same year that our king, King George V had his Silver Jubilee in 1935.

The Silver Jubilee of George V was on 6 May 1935. It marked 25 years of his reign as the King of the United Kingdom, British Dominions, and Emperor of India. The Jubilee was designed to lift the spirits of a country in depression and was marked with events around the country.

It was the first ever Silver Jubilee celebration of a British monarch. The Singer Company cleverly took this on board and presented their latest and finest masterpiece as the Jubilee Sewing Machine.

However long before its official release in 1935 work had begun on the Singer 201.

New machines don't just appear on the shop shelves all over the world. Years of research and development has taken place. Years of trial and error, of plants setting up foundry castings, engineers perfecting the designer's ideas until finally, a machine is born.

After years in development the Singer 201 had finally arrived. Notice the wonderful clearance under the arm of the machine. With its narrow foot and high clearance the 201 soon became the dressmakers dream. When quilting took on in a big way as a hobby, quilters sought out the electric cast iron 201s. Once again the high clearance and a narrow foot, with a superb feed was what quilters loved. Today the 201 still out performs 99% of all sewing machines when piecing and quilting. All you had to do was drop a motor on and away you went.

Interestingly the Elizabethport 201's were not up to the same quality of castings as the Kilbowie plant. You can always tell an unmarked 201 from the American plant by the rougher internal casting. No letters please!

This is the British Singer 201K-4 cast iron hand crank. It came in at a hefty 36lbs but like all professional machines, you don't take them shopping. You pop them on the sewing room table and hammer them. It was a quick job to replace the hand crank with a motor if needed.

In the case of the Singer 201 the first major problem was the cost of manufacturing. All the precision parts added up. While it may give the opposition ulcers trying to copy the ideas it also cost a fortune.

It is estimated that the Singer 201 was the most expensive domestic machine that Singers had developed up until 1930. I've never found out the costs but it must have been close to building a rocket for space back in that period.

When such a huge investment is put in, then the launch needs to be just as spectacular. Somewhere in the development of the Singer 201, designers and engineers must have realised that they had something very special. The silky smooth quietness of the machine had never been seen in a lockstitch before. The closest machine before that would have been the Willcox & Gibbs silent chain stitch. Now for the first time Singer had a winner of unprecedented proportions.

PRICE LIST

Treadle and Cabinet Table Machines

201K Machine on One-Drawer, Drop-Leaf Table and Cover	List Price £17 15 0
If paid at 12/- per month—Less 5%	£16 17 3
" " 20/- " — " 10%	£15 19 6
" within 3 months— " 15%	£15 1 9
Net Cash Price— " 20%	£14 4 0
201K Machine on Three-Drawer Cabinet Table	List Price £20 5 0
If paid at 12/- per month—Less 5%	£19 4 9
" " 20/- " — " 10%	£18 4 6
" within 3 months— " 15%	£17 4 3
Net Cash Price— " 20%	£16 4 0
201K Machine on Five-Drawer Cabinet Table	List Price £21 5 0
If paid at 16/- per month—Less 5%	£20 3 9
" " 20/- " — " 10%	£19 2 6
" within 3 months— " 15%	£18 1 3
Net Cash Price— " 20%	£17 0 0
201K Machine on No. 46 Enclosed Cabinet	List Price £25 10 0
If paid at 16/- per month—Less 5%	£24 4 6
" " 20/- " — " 10%	£22 19 0
" within 3 months— " 15%	£21 13 6
Net Cash Price— " 20%	£20 8 0

The 1935 price lists show the cost of the British Singer 201 and the various cabinets available.

PRICE LIST

Electrically Driven Machines

201K Portable Machine, with Wood Base and Cover, Built-in Motor (K2) or B.R.K. Motor (K3), Extension Table and Singerlight	List Price £20 10 0
If paid at 12/- per month—Less 5%	£19 9 6
" " 20/- " — " 10%	£18 9 0
" within 3 months— " 15%	£17 8 6
Net Cash Price— " 20%	£16 8 0

201K Machine on No. 40 Table, with Built-in Motor (K2) or B.R.K. Motor (K3) and Singerlight	List Price £32 12 6
If paid at 16/- per month—Less 5%	£30 19 10
" " 20/- " — " 10%	£29 7 3
" within 3 months— " 15%	£27 14 7
Net Cash Price— " 20%	£26 2 0

Hand Machine

201K Machine with Base, Cover and Extension Table	List Price £13 17 6
If paid at 12/- per month—Less 5%	£13 3 8
" " 20/- " — " 10%	£12 9 9
" within 3 months— " 15%	£11 16 11
Net Cash Price— " 20%	£11 2 0

Note how Singer called their electric 201 Built-in motor. We now call this direct drive motor a potted motor, 201K-K2. The 201K-K3 was the belt drive version. This was by far the most popular version in Britain. Mainly because stockist and Singer Dealers could easily swap between electric, hand or treadle when a customer ordered a machine. One machine to stock with three options.

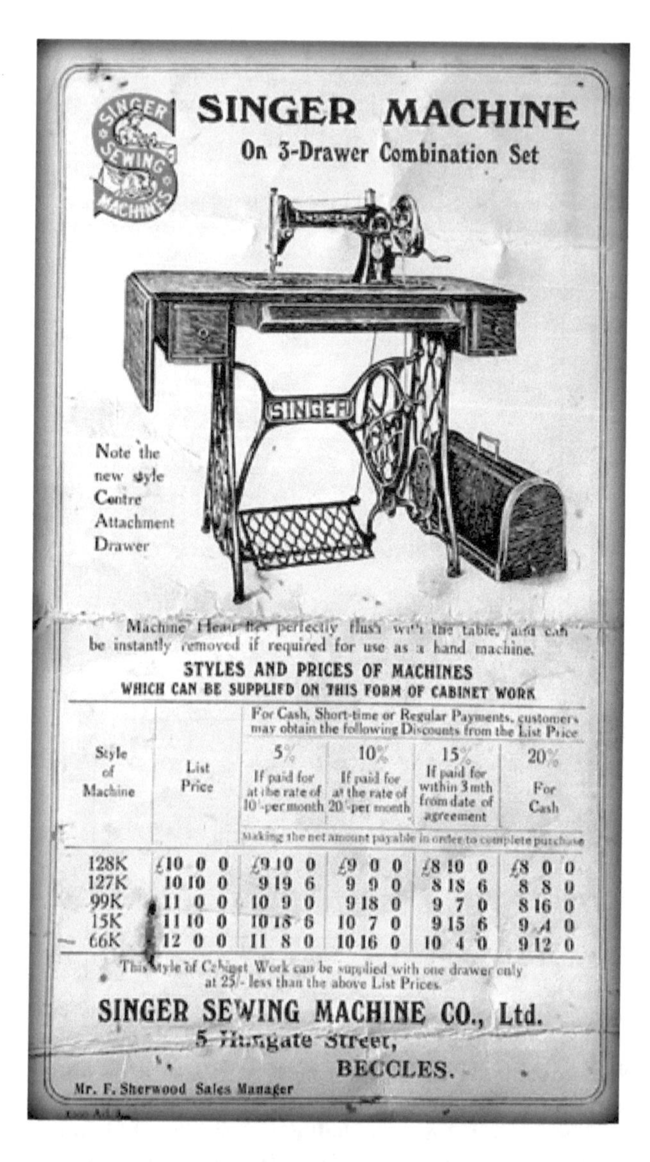

SINGER SEWING MACHINES

SINGER MACHINE

On 3-Drawer Combination Set

SINGER

Note the new style Centre Attachment Drawer

Machine Head lies perfectly flush with the table, and can be instantly removed if required for use as a hand machine.

STYLES AND PRICES OF MACHINES

WHICH CAN BE SUPPLIED ON THIS FORM OF CABINET WORK

Style of Machine	List Price	For Cash, Short-time or Regular Payments, customers may obtain the following Discounts from the List Price			
		5% If paid for at the rate of 10/- per month	10% If paid for at the rate of 20/- per month	15% If paid for within 3 mth from date of agreement	20% For Cash
		Making the net amount payable in order to complete purchase			
128K	£10 0 0	£9 10 0	£9 0 0	£8 10 0	£8 0 0
127K	10 10 0	9 19 6	9 9 0	8 18 6	8 8 0
99K	11 0 0	10 9 0	9 18 0	9 7 0	8 16 0
15K	11 10 0	10 18 6	10 7 0	9 15 6	9 4 0
66K	12 0 0	11 8 0	10 16 0	10 4 0	9 12 0

This style of Cabinet Work can be supplied with one drawer only at 25/- less than the above List Prices.

SINGER SEWING MACHINE CO., Ltd.

5 Hungate Street,

BECCLES.

Mr. F. Sherwood Sales Manager

Here you can see Singers old baby the Singer 66k. Also note the prices of the older models. Most of these were available right back in the Victorian Era!

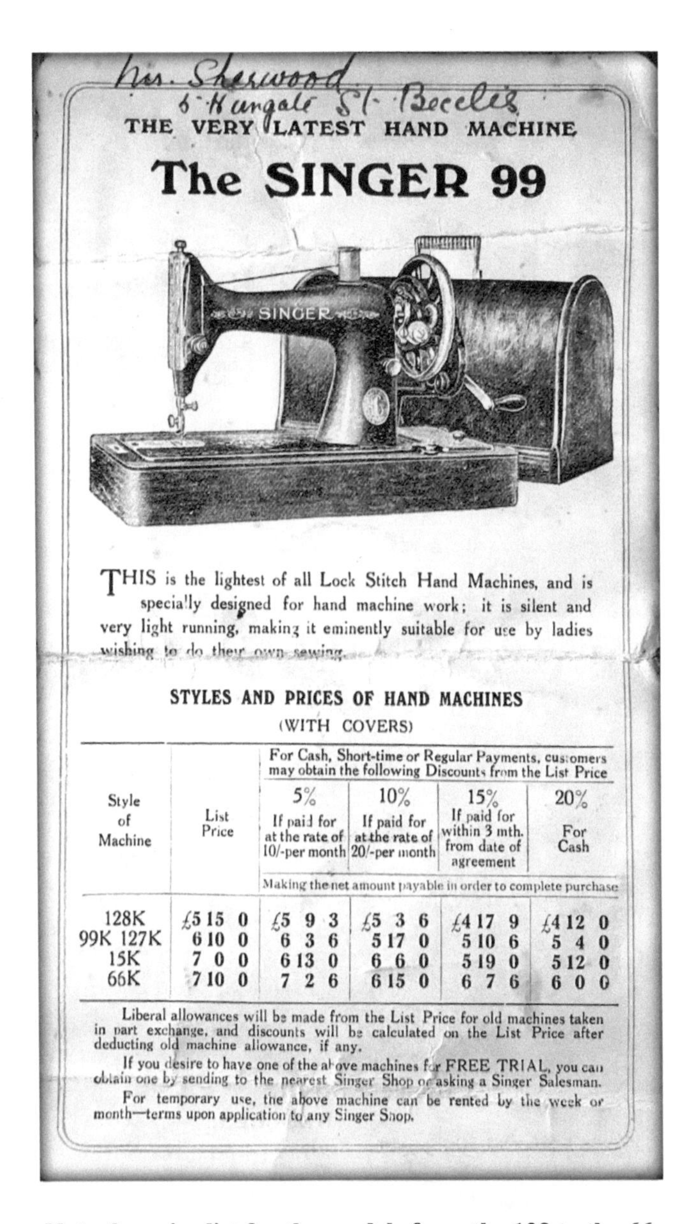

THE VERY LATEST HAND MACHINE

The SINGER 99

THIS is the lightest of all Lock Stitch Hand Machines, and is specially designed for hand machine work; it is silent and very light running, making it eminently suitable for use by ladies wishing to do their own sewing.

STYLES AND PRICES OF HAND MACHINES

(WITH COVERS)

Style of Machine	List Price	For Cash, Short-time or Regular Payments, customers may obtain the following Discounts from the List Price			
		5% If paid for at the rate of 10/-per month	10% If paid for at the rate of 20/-per month	15% If paid for within 3 mth. from date of agreement	20% For Cash
		Making the net amount payable in order to complete purchase			
128K	£5 15 0	£5 9 3	£5 3 6	£4 17 9	£4 12 0
99K 127K	6 10 0	6 3 6	5 17 0	5 10 6	5 4 0
15K	7 0 0	6 13 0	6 6 0	5 19 0	5 12 0
66K	7 10 0	7 2 6	6 15 0	6 7 6	6 0 0

Liberal allowances will be made from the List Price for old machines taken in part exchange, and discounts will be calculated on the List Price after deducting old machine allowance, if any.

If you desire to have one of the above machines for FREE TRIAL, you can obtain one by sending to the nearest Singer Shop or asking a Singer Salesman.

For temporary use, the above machine can be rented by the week or month—terms upon application to any Singer Shop.

Note the price list for the models from the 128 to the 66. These were all the 201's competitors made by Singer and doing a similar job. Amazingly the Singer 99 was the lightest machine in the range and weighed the same as a baby elephant!

Notice the spoked hand wheel above and the solid below. This difference is a big pointer to which model of 201 you may have. Note that you put the cable into the light socket as most houses in the 1930s didn't have plug sockets.

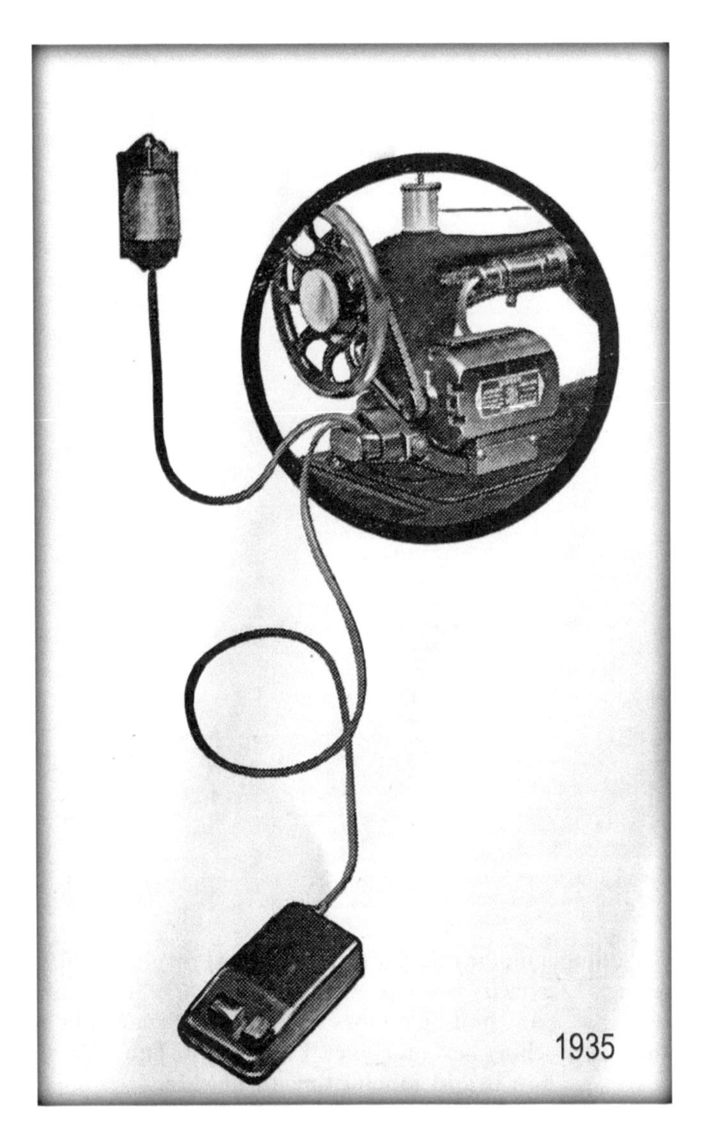

Here you can see the complete bolt on 201-3 which took a B.A.K. OR B.Z.K. motor. Once again power came from the light in the room so most motors came complete with their own light unit for the machine.

Adding a motor and light was the latest revolution in sewing. Electricity was sweeping the world and Singer was selling as many bolt-on motors for their older machines as they were selling new machines. I have a YouTube clip on how to add a motor to your machine.

Sewing machine cabinets were big business. Apparently the Singer factory in Scotland employed over 2,000 cabinet makers at its height. Styles of sewing cabinets changed with the times as did the hairstyles. Before Hollywood really took off it was the 'Royals' who filled the pages of the press and if Princess Elizabeth changed her hair, so did many others.

THE NEW No. 40 TABLE

with 201K2 or 201K3 Electric complete

THE 201K Electric is operated in the utmost comfort when fitted on this Table, as both hands are free to guide the work.

With the "leaves" folded back and the head of the machine in sewing position, it is only necessary to plug into an electric lamp socket or wall-plug to transform the table into an efficient all-electric Sewing Machine.

Closed, and with the starting lever turned up out of sight, the table is a useful and artistic piece of furniture.

The 15K82 and 66K6 Machines can also be supplied on this Cabinet Work.

How do you sell a machine that cost £20.10 in 1935? The price was much more than their other domestic machines. The Singer 66 doesn't look that different in shape and size. The Singer 15 was so strong that demonstrators used to sew through tin cans to impress the bored husbands accompanying their wives at the Singer shops. I know that the 15K will do that because I did a YouTube clip on it (it's still floating around somewhere).

Singer girls were trained at the London top stores on how to use the 201 and all its attachments. They would pass on this information to new buyers. Free lessons were included with every purchase.

The Singer 66 was old-world stunning and came from another era. Decorated in Egyptian garb the 66 reflected the period of Tutankhamun. American decoration was called Red-Eye. The 66 had no drop feed or reverse. It looked similar to the 201 but was vastly different.

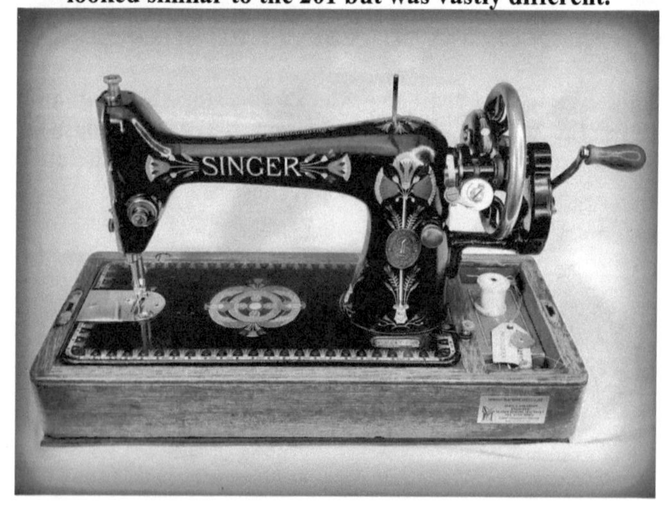

The Singer 15 was Singers toughest domestic and, as it turned out, its longest running machine. The machine was in production from the late Victorian Period up until the original Singer Company closed. In later years the Singer 15 machines came from the Far East and Asia. They were rough and tough but boy they were indestructible. In later years they added a reverse and drop feed but little else changed. The Singer model 15 is the most widely copied sewing machine of all time. It is still made by several makers to this day. Its simplicity in construction and low cost meant that it outlasted just about every Singer model ever made.

Three shillings a week was the perfect way for the Singer Company to attract new buyers to their super 201. So for three shillings a week you could buy the best domestic straight stitch sewing machine on the planet.

Besides free tutorials on the new 201, Singers secret was to keep to their low weekly payments but just extend the period that a customer paid for their machine. I met a woman once who bought her machine new in 1926 and finished paying for it in 1941. She even saved it from a Nazi bombing raid. That story is in one of my On The Road Series and it's a blinder. Fifteen years of payments for a sewing machine, WOW!

Attachments

Singer 201 machines came with a standard set of attachments but local Singer shops could always throw a couple more in to close a deal. These little extras often go unused but all 201 machines did come with free training on them.

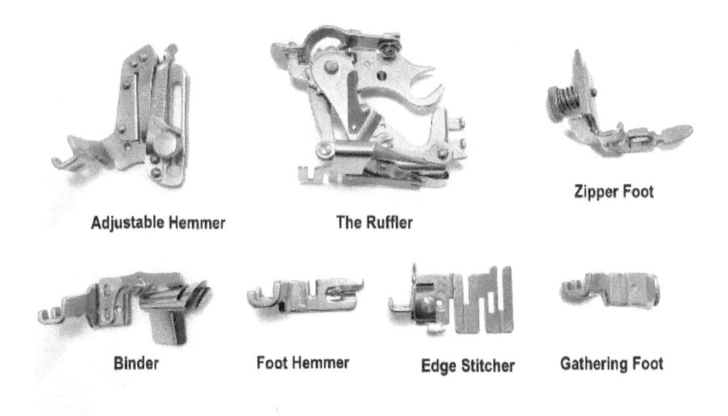

As well as instructions, oil, screwdrivers, needles and bobbins the usual standard attachments were…

1. Standard hemming foot.
2. Large adjustable hemming foot.

3. Zip foot.
4. Gathering foot.
5. Bias binding foot (depending on the fashion).
6. Ruffler and pleater.

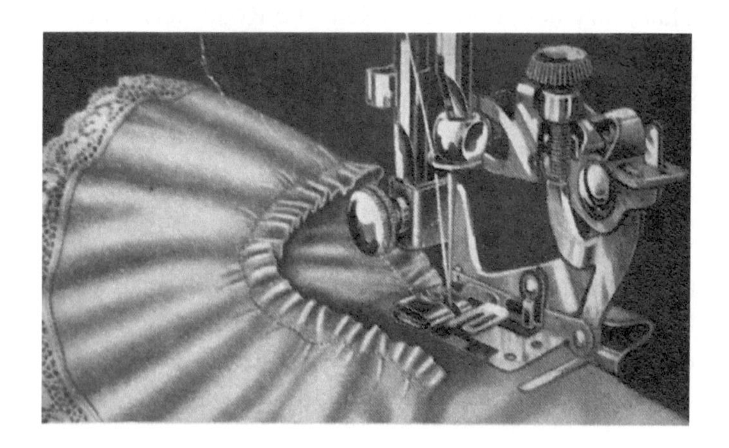

The pleater was an amazing bit of kit and in the right hands could work wonders.

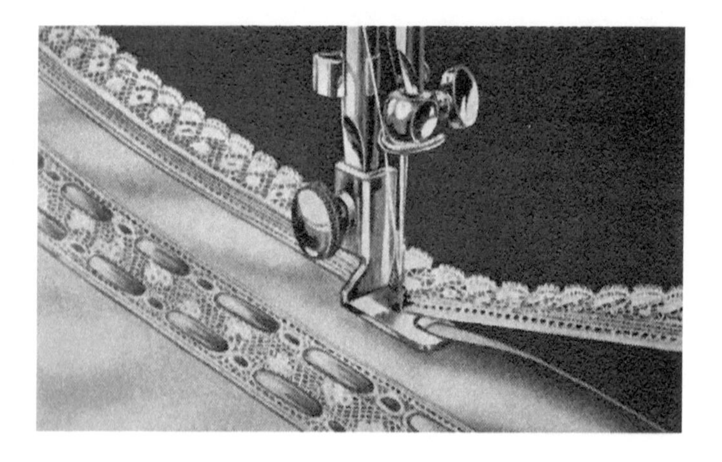

Hemming on the 201 was a dream and a tablecloth or skirt hem could be sorted in moments with the right foot.

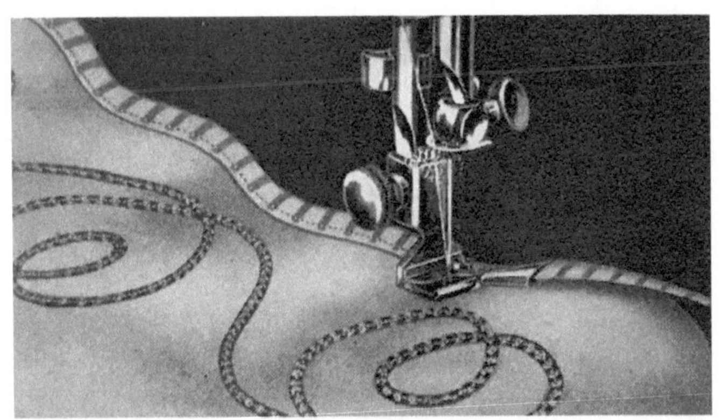

There were many other feet from piping feet to assorted shirring, quilting and braiding feet that could always be purchased. Each came with its own instructions.

Because the 201 was a standard LOW-SHANK fitting, attachments for the 201 were vast, from automatic buttonholers to zigzag and pattern making feet. If you wanted or needed a special foot, there was usually one to help. Funnily the zigzag foot worked by wiggling the work (as the needle could only go up and down). Basically the work wobbled its way through the machine like a drunken sailor.

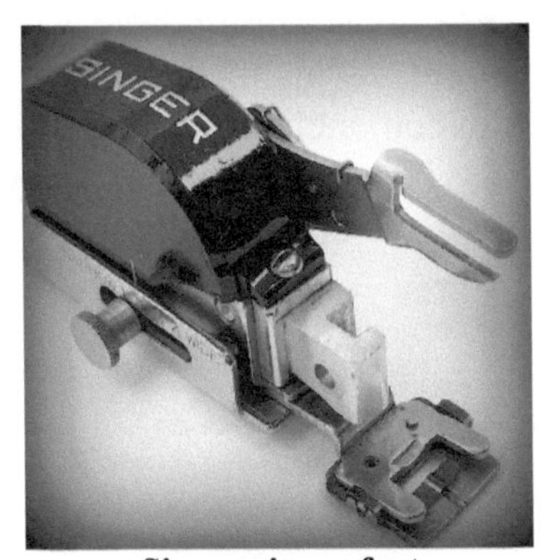

Singer zigzag foot

Each Singer shop sold an assortment of threads and materials too. They always promoted regular servicing of Singer machines. Singer dominated the high streets of Britain having hundreds of shops and thousands of employees. A trip to any Singer shop was a joy, full of fabrics and fashions.

In our new era of instant information online, for every one of the feet Singer made available, there is a wonderful YouTube clip, showing you how to use them and what work you can create.

There are countless feet and countless YouTube clips. YouTube has become the teacher in your home. You can find out just about how to do anything from build a rocket to using your sewing machine. I currently have over 100 YouTube clips online showing my collection and sewing tips.

You can also look up several clips I have on the Singer 221 and 201. My Singer 201 Demonstration had over 40,000 views last time I checked.

Also I have some great clips on Free Motion Work and threading your sewing machine. My later books seem to work in conjunction with my site www.sewalot.com, and the Internet. They are the better for it. Look me up and enjoy.

SINGER SERVICE

Wherever you go you will find expert, dependable **SINGER*** Service nearby. SINGER is interested in nelping you keep your SINGER Sewing Machine in top condition. That is why you should always call your SINGER SEWING CENTRE if your machine ever requires adjustment or repair. When you call your SINGER SEWING CENTRE you can be sure of obtaining the service of a trained **SINGER** repair man and can be assured of warranted **SINGER*** parts when needed! Look for the familiar red "S" Trade Mark on your SINGER SEWING CENTRE and the ever ready SINGER Service Car.

EVERYTHING FOR THE WOMAN WHO SEWS

The answer to your sewing needs is at your SINGER SEWING CENTRE. There you will find a wide choice of sewing necessities and notions, sewing instructions and guidance and services for covering buttons, hemstitching, making belts and buckles, to mention a few. Look in your telephone directory under SINGER SEWING MACHINE COMPANY for the SINGER SEWING CENTRE nearest you.

Anyone who sewed would visit one of the hundreds of Singer shops. The London stores were the largest sometimes having over 20 staff.

WHEN YOU OWN A SINGER 201K MACHINE

you have a beautifully styled, smooth running machine which offers you a new world of sewing enjoyment. Exclusive dresses for yourself, clothing for your family, and numerous items for the home—all will be yours at a fraction of their ready-made cost.

TO GET THE MOST ENJOYMENT FROM YOUR SINGER

You are entitled to sewing lessons when you become the owner of a new SINGER. A skilled, SINGER-trained teacher personally guides you and assists you in learning the fundamentals of home sewing. Other courses embracing all phases of home sewing are available at low cost.

Strangely, less than 50% of customers who purchased new Singer 201 machines took up the free offer of lessons. That could be because they knew how to use most of the attachments already. Singer customers were loyal customers.

CHAPTER 2
Singer 201 Models

By 1935, the Singer 201 was ready for its 'official' premier. Pre 1935 test 201's had been a big hit with everyone that used them. The sales and advertising was ready, including training for the Singer staff (for the free lessons that went with the machine). Now, let us look at 201 models from 1935 to 1962.

I often feel like Howard Carter searching in the Valley of the Kings forever looking for that one last piece of the puzzle. With Singer 201 models I have found out lots, some from manuals printed over the decades from 1935 to 1962. Some from the countless advertising leaflets I have acquired. Some from old retired Singer men (who often have the best tales). I search libraries, patent offices, my god I was once caught scrubbing moss from Isaac Singer's grave down in Paignton, Devon. So here is what I have found. It will be for other enthusiasts to complete my work.

Don't forget many years have gone by since these machines left the dealers and agent's shops. Many machines were modified to specific customers' requirements, motors switched to hand cranks, lights added or removed and so on. Boxes and cabinets were switched by dealers, new machines dropped into old units. So this list is not gospel but more of a general guide.

The Singer 201 (often but not always) came with a few letters such as the 201k (the K that sometimes follows the number denotes that it was made in Kilbowie, Scotland) or the Singer 201P which was assembled in Australia using parts from Kilbowie. The Singer 201D with the American style potted motor was from Wittenberge in Germany.

There were also subclasses such as Singer 201k1 or 2 or 21 etc. They were pretty much identical with minor alterations. For example I've been told by an old Singer agent that the 201-6 came with a knee control and light. The 201-7 had a knee control and special motor bracket.

Then there were the special conversions from treadle foot-power to electric. A modified foot controller would be bolted to the frame of an old treadle plate so that you could still use your original treadle but now, as you pressed on the foot plate, it electrically sewed. I've never found out that subclass number.

Singer Bracket Universal Electric Motor.

Note how you plug the power into the light socket.

The 201-13 subclass which you will probably never see listed anywhere, was electric but with a knee bar speed controller rather than a foot pedal. This was popular after WW2 with a high number of casualties unable to use a foot pedal. Some were even designed to be used with the knee rod taken out of the wood base and turned up to be used by an operators elbow.

The 201-15 was a knee & arm control but also had a speed reducing bracket for embroidery work. So as you can see there were many subclasses but they were all minor modifications. For the three decades that the 201 spanned, there were hardly any major modifications. It was as if the Singer Company had invented the shark, a perfect evolution of a perfect stitch. It could have little improvement.

Now the big change in Britain came in 1954 with the addition of the aluminium Singer 201. After WW2 Singers had managed to secure enough alloy to make the 201 a little lighter.

Depending on the case and base the cast iron 201 weighed around 36lbs, sometimes up to 40lbs depending on the goodies and box. The later alloy ones were around 23lbs on their own. Interestingly, a Singer featherweight, in its case and with accessories, is about six pounds lighter than an alloy 201 without its case. The Featherweight alone is only 11lbs and 17lbs complete in its box.

Now, if Singer put the alloy 201 in a lightweight base it would knock several pounds more off in weight. However, that weight was essential when pushing work through the machine. A common problem with very light machines (besides not being able to sew heavy work) is that they constantly move away from the operator when sewing. Twenty three pounds was considered just about right.

They still continued with the heavier cast iron or pig iron models as well. Many people loved the greater clearance on the iron models.

The alloy Singers were only available from Kilbowie. They were lighter and still as smooth as silk. However the wider body lost some of its clearance under the arm. Now for a little more detail.

Germany 1935-1944

Let's look at the smaller manufacturing outposts first. Wittenberge (later Wittenberg), Germany.

From 1935 (maybe earlier) until its dramatic takeover by the Russians late in WW2. The Wittenberge Factory in Germany produced the Singer 201 in hand, treadle and electric, all in cast iron. These were marked (sometimes) 201D. They did not produce many machines as they helped with the war effort from the late 1930s. Often the German 201 had no D prefix which is understandable as consumers in different countries in the 1930s were not buying German goods.

Here is the German Singer 201D sewing machine made in Wittenberge. Potted motor, integral front light circa 1940.

It is interesting to note that economies were being made even on this model. Note the cheap transfer badge that was on the last German 201D sewing machine model, compared to the earlier brass plate below and on the British 201s.

This is an earlier badge from a 201D made in Wittenberge, Germany. Some new evidence coming to light suggests that engineers at the Germany factory actually helped to develop the Singer 201. The Wittenberge Singers were not always stamped 201 and a few may even pre-date the official 1935 launch date.

Australia
1956-59

I've been told many times that Australia made no Singer 201s at all! However, from around 1959 until 1961 they assembled alloy Singers sent over in parts from Kilbowie in Scotland to Penrith in NSW, Australia. These were marked (sometimes) with Singer 201P. To my knowledge these were all aluminium only, electric, and all tan beige brown models. Some were converted to hand by suppliers. This is an easy job for a specialist which entails swapping the solid hand wheel for a spoked hand wheel and winder assembly.

The Singer Penrith Manufacturing Company also put together several other Singer models with parts supplied from Singers European plants. 210P-21 treadle, 201P-23 belt drive, bolt on motor. 201P-24 hand crank. No records have come to light from the Penrith and may have been destroyed.

You can assume that importing the machines in pieces and assembling them at Penrith was for import tax reasons. It was a good way of supplying machines and avoiding preventable duty. Nearly all the major manufacturers tried to keep their costs to customers down by legally avoiding unnecessary taxation. Novum had a wonderful 15K style machine that was 'made in Ireland'. No one ever found the factory!

North America

Out of all the plants in North America, from St. Jean-sur-Richelieu in Canada, to Bridgeport in Connecticut, South Bend, Indiana, and Anderson in Carolina, only the Elizabethport factory in New Jersey produced the Singer 201s.

It is very possible that some of Singers other plants produced parts for the 201 but only Elizabethport made complete 201s, AND ONLY cast iron or pig iron models were made there.

In all my research and constant questioning, to my knowledge NO alloy Singer 201s were ever made in North America. Of course plenty turn up but they all would have been imported. Alloy 201s are rare machines in the US.

America
1935 to 1957

201-1 Treadle, spoked wheel (optional Hand Crank).
201-2 Geared 'potted' motor (solid hand wheel). These models also often had integral front lights rather than bolt on rear lights.
201-3 External bolt-on motor with belt (solid or spoked wheel).
201-4 Hand Crank spoked wheel.

Great Britain
1935-1962

In the United Kingdom the factory at Kilbowie, on the banks of the River Clyde in Scotland, produced by far the largest number of Singer 201s with many options.

201K-1 Treadle, spoked wheel (optional Hand Crank).
201K-2 Geared 'potted' motor (solid hand wheel). These models also had integral front lights rather than bolt on rear lights.
201K-3 External bolt-on motor with belt (solid or spoked wheel).
201K-4 Hand Crank spoked wheel.

The Singer 201K-4 spoked wheel, hand crank was the biggest seller in the UK. Smaller homes and cramped spaces meant that a hand crank could easily be slipped away when needed. Hand cranks outsold treadles by 4-1. I have been told in America it was the other way around. Motors were later added.

I have a true tale in one of my 'On The Road Series' of books. It's about Singers supplying a hand crank to a customer with only one arm! She had brought up a family of four children on her own. The company failed to take into account that the old Singer they were replacing (in front of the press) was a treadle, not hand crank (which she could use with one arm). It caused a national incident at the time.

DO YOU KNOW
THAT . . . . LITTLE
SINGER MOTOR?

One screw attaches it to your machine

Cuts out all fatigue!
Makes sewing a pleasure!
Better work done in half the time!

And then—after dark, that Little

SINGERLIGHT

—by showing the stitching more clearly—prevents eyestrain and saves time and annoyance when threading the needle

Any Singer Shop or Singer Salesman will tell you all about them

Singer lights were often sold separately, as a bolt on optional extra, bought as a motor and light package, or supplied with the machine complete. Each method came with a different subclass reference.

Also electricity was much slower spreading across the UK and was unreliable. Many homes stuck with 'human powered' Singers. That is why many more hand cranks turn up in the UK than almost anywhere in the world. Many children's first go on a sewing machine was by hand crank and they are still popular today.

It is reported that it was in Scotland that the Singer craftsmen perfected the steaming of the wood to create the bentwood cases for the machines. Isn't this just the perfect scene of tranquil calm!

Alloy Singers in Great Britain 1954-1962

Introduced in 1954 from surplus alloys, the lightweight Singer 201K series were a huge hit and sales boomed.

201K-21. Alloy machine in a treadle base.
201K-23. Alloy with bolt on motor, belt drive and rear light unit. These were occasionally exported to America and Canada with different voltage motors.
201K-24. Some refer to this as the 201K-29 Black alloy hand crank with spoked cast iron hand/balance wheels. These are the rarest of all Singer 201s.

At some point during the manufacture of the alloy Singer 201s in Britain they started producing more aluminium models than iron. This may have coincided with the change in colour from black to brown/beige in 1955 but I am not sure.

One final point. The pig iron Singer 201s weighed around 36lbs and the alloy-aluminium ones 23lbs. Now, total weight depended on the bases and of course the cases (there were several). The 201s were supplied with bentwood cases originally in the 1930s and finally in the UK 'mock-croc' cases and deluxe hinged or drop-on light fabric covered beech-wood cases. New plastic cases are available today.

Now that we have dealt with all the main models our next chapter will deal with a little know oddity in the 201 stables, the Singer 1200.

One of the many options available was the electric knee speed control instead of the foot control. The foot controller was actually tucked into the base of the machine and operated by bars and rods that led to the knee lever.

CHAPTER 3
The Singer 1200

Later we are going to talk about the Singer 101, forerunner to the 201, but now we are going to talk about another stablemate to the 201, the Singer 1200. Someone at the Elizabethport factory had a brainwave.

The Singer 201 had proved it was a workhorse. It had as much and more clearance than many industrial machines and was capable of high speed sewing. It could also handle just about anything that you could push through it.

Electrically powered industrial sewing machines usually ran with massive, heavy and noisy motors strapped under the industrial tables. These induction motors sounded like aircraft taking off. In a factory situation on an estate, noise is not such a problem. However, put one of these industrial machines in a flat or housing estate and the whole place could shake. This led to countless complaints if machinists sewed late at night or early morning.

Now, I grew up in that industry and we had machinists sewing in their own homes as well as the factory machinists. It was a common problem for one of our workers to get a visit from the local council about noise disturbance from their industrial machine. So, how about taking a Singer 201 and adding a few features for the home machinist?

At the Elizabethport factory they mounted the 201 into a neat table so that it had a large flat working area. They added their best direct drive, 'potted motor'. They then added a knee lift so that, just like their industrial machines, the operator could use both hands to hold the work while lifting the presser foot with their knee. Oh, quilters were going to love that. The presser bar also raised the sewing foot a little higher than the normal 201. An integral front fitting light gave the best vision and a host of attachments added the final touch.

The addition of the knee lifter to the 201-2 and a new model number (1200) gave the 201 a new market. Singer shops could sell this work horse with confidence to professional and light industrial machinists. When a model 1200 does turn up it is usually hammered. You can see by the countless pin scratches on the back of this one owned by Lori Thomas that the machine has earnt its pay.

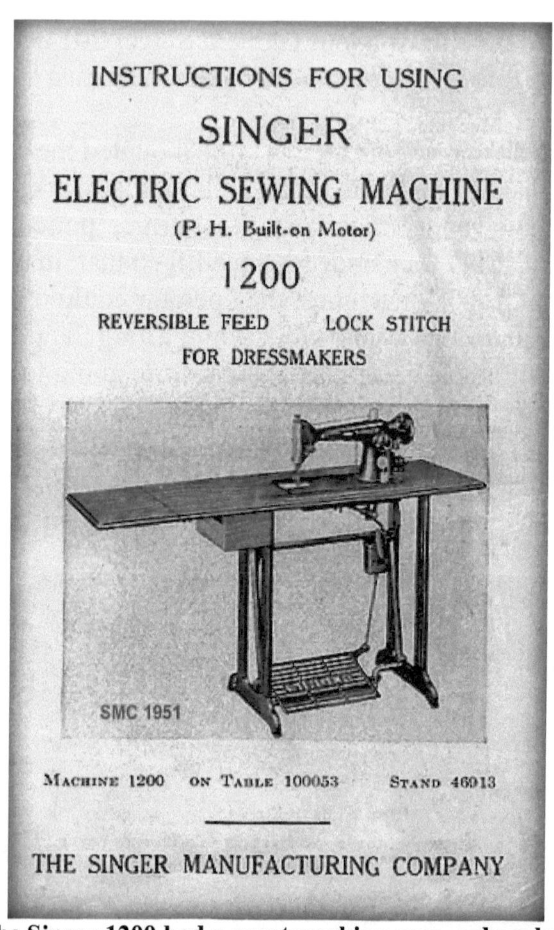
INSTRUCTIONS FOR USING

SINGER

ELECTRIC SEWING MACHINE

(P. H. Built-on Motor)

1200

REVERSIBLE FEED LOCK STITCH

FOR DRESSMAKERS

SMC 1951

MACHINE 1200 ON TABLE 100053 STAND 46913

THE SINGER MANUFACTURING COMPANY

The Singer 1200 had a great working area, a drop leaf extension and a large drawer for all the extra attachments. The purpose made base (modified from treadle parts) was light and easy to use and included an oil drip tray with knee lifter. The 1200 was first called the 1200-1.

A comprehensive instruction manual was added. It included all the maintenance and adjustments as well as instructions on each of the extra feet.

And so the Singer 1200 was born. Initially called the 1200-1 as Singers assumed many more models would soon follow. The ideal dressmakers and homeworkers machine was born. Smooth, silent, reliable and virtually indestructible. Rather than push their machine as a full industrial, they leaned towards professional seamstresses and machinists, wedding dress, bridesmaids, curtain makers and dressmakers. Alteration specialists and a host of artisan homeworkers all bought the machine.

The Singer 1200

Here is a Singer 1200-1 in all her glory, kindly sent in by Lori Thomas. You can clearly see the knee lift (also used by the Singer 31 machines). It looks so simple but just the knee lift required nearly 30 extra parts through the entire machine.

I believe the first sales of the Singer 1200 in Britain was in Singers Centenary year of 1951. To my knowledge the modified 201's were only made at the Elizabethport factory in America. From the figures that have come to light so far it seems to pre-date the 1935 Singer 201 launch by a year and may have been available in America from 1934 onward, and in Britain from 1951. World War Two is the probable reason for this. One day new figures may come to light but for now we stick to the manufacturing dates for the Singer 1200 as 1934-1956.

Total numbers of the 1200 produced were very low. Estimates put it at between 6-10,000 units. There is a major problem with the serial numbers. Basically the Singer 1200 was a modified Singer 201-2 with extras. Because of this a serial number on a Singer 1200 often comes with the correct year of manufacture but as a 201 model not 1200.

To identify if you have a Singer 1200 with no markings look at the back of the machine and check for the knee lifter mechanism. Even if it has been removed there will be tell-tale signs, holes and extra parts around the presser lifter area and bed.

Note the 1200 badge on the modified Singer 201-2

So why did the 1200 go out of production when it was the ideal machine? There was a problem with the Singer 1200. It was too good at its job. If you were a manufacturer, selling industrial and domestic machines and came up with one machine that did both, what would happen? Of course, you would drop sales.

Okay, the 1200 was no heavyweight industrial but it was extremely good at what it did. Often 1200s turn up heavily worn from decades of hammering. They say if you could squeeze your work under the foot, the machine would sew it!

So, by 1956 apparently the 1200 was no more and Singers went back to selling mainly industrial AND domestic sewing machines.

Now that we finally have our confusing models out of the way, let's step back a moment and look at the period in which the unique 201 machine was first developed.

It is a fascinating part of our social history and part of the Singer 201 story.

CHAPTER 4
A Little History

Cheap always sells, but only quality lasts...
After the Great War, mass production from technologies learnt during wartime, and the supply of electricity into houses led to massive advances in sewing machine technology. The pinnacle of this technology by the 1930s can be seen in the stunning performance of Singers 1935 offering, the formidable model 201, which was produced in Britain, Germany and America.

This masterpiece of engineering was expensive but opened up a whole new era of opportunities amongst the working classes of Britain.

The 1930s were hard, the world was suffering from a massive depression sparked by the Wall Street Crash in America. The ripples were felt around the globe and millions of people found themselves out of work.

In 1926 future Prime Minister Stanley Baldwin put forward legislation to create a 'National Grid', allowing electricity to reach around the United Kingdom.

A rare 'badged' Singer 201K4 (K4 cast iron hand crank) kindly sent in by Graham Reeve. I had not seen a marked K4 before, normally they were just stamped 201K. As you can see by the open-spoke wheel, this one was originally supplied as a hand machine. The 201K-2 or 201-2 below often had a front integral light unit along with a direct drive motor behind. The front lights got very warm but with new LED bulbs that's not a problem today.

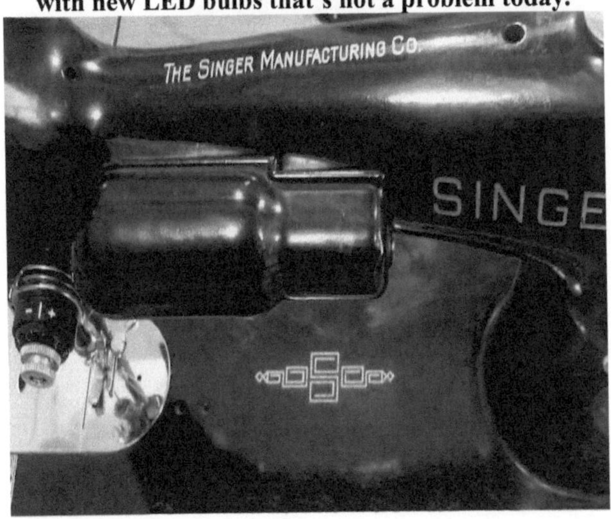

The early back and side plates of the first 201s were beautifully ornate with a real Art Deco touch. The machine oozed the 1930s.

Later designs were either painted or striated.

Later alloy models were a smart 1950s modern two-tone brown. New era, new style. Inside the machines were practically identical.

Here you can see the B.A.K. 4-12 bolt-on belted motor with light. You can see that the back panel of the sewing machine can be removed for oiling. There is one very important oil point on the main vertical shaft that is only accessible through the plate.

Possibly the rarest of all Singer 201s The Texas Centennial exposition Singer 201 from 1936.

In Britain, unemployment and poverty manifested itself in uprisings like the Jarrow March in October of 1936 (Google it, it's fascinating). However new opportunities were also opening up. For the first time a truly professional machine was now available to the masses. There was almost a Singer shop in every town and city in the country and in the shop windows was the new Singer model 201.

Although expensive, with the help of hire purchase, part payment or layaway scheme, first introduced on mass by Singer and his partner Edward Clark in the 1850s, women found that they could buy the Singer 201 and pay for it over a period of years, while earning a living from the machine at the same time. Women, who had returned to housework after WW1 were ready to take up the challenge once more with a machine that could handle professional quantities of work without fuss. They would go on to earn a living and help with the household expenses again.

The San Francisco Golden Gate Exposition of 1940 is another super-rare Singer badge (and other years too).

Super quality construction and simplicity in design gave the Singer 201 amazing performance, durability and a seemingly indefinite lifespan.

Oiling the Singer 201

Fig. 13.

After oiling, run the machine rapidly for a few minutes, so that the oil may reach the bearings. If in constant use, the machine should be oiled two or three times a week. **Neglect to do this will shorten the life of your machine and cause you trouble and annoyance.**

To oil the stand, apply a drop of oil to the centres upon which the band wheel and treadle work, and to both ends of the pitman rod which connect the treadle with the band wheel.

Fig. 14.

I have a YouTube clip on oiling and servicing your machine, 'Alex Askaroff Presents, servicing an antique, vintage sewing machine'. Check it out.

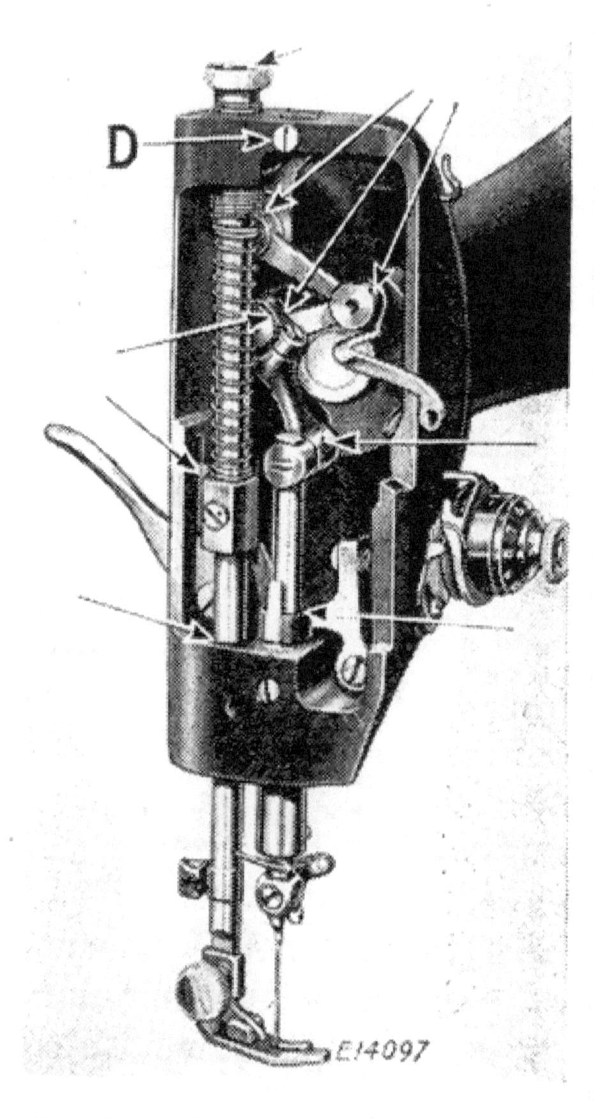

The best oil to use is a sewing machine oil which is a high grade non-staining mineral oil which never hardens. Gun oil has similar properties. I have found one drop in the oiling spots once a month is fine. Over oiling can lead to oil on your work. D is the side plate cover screw.

With regular oiling and maintenance the Singer 201 work horse could be used hard and almost continuously. Whether making curtains or costumes machinists could earn a living with the 201.

For normal domestic use the Singer 201 would have been supplied as a hand machine (201-4) or treadle but for the professional it could have a motor attached to it that could run at over 1,100 stitches per minute.

The new 201 helped make all the latest 1935 fashions.

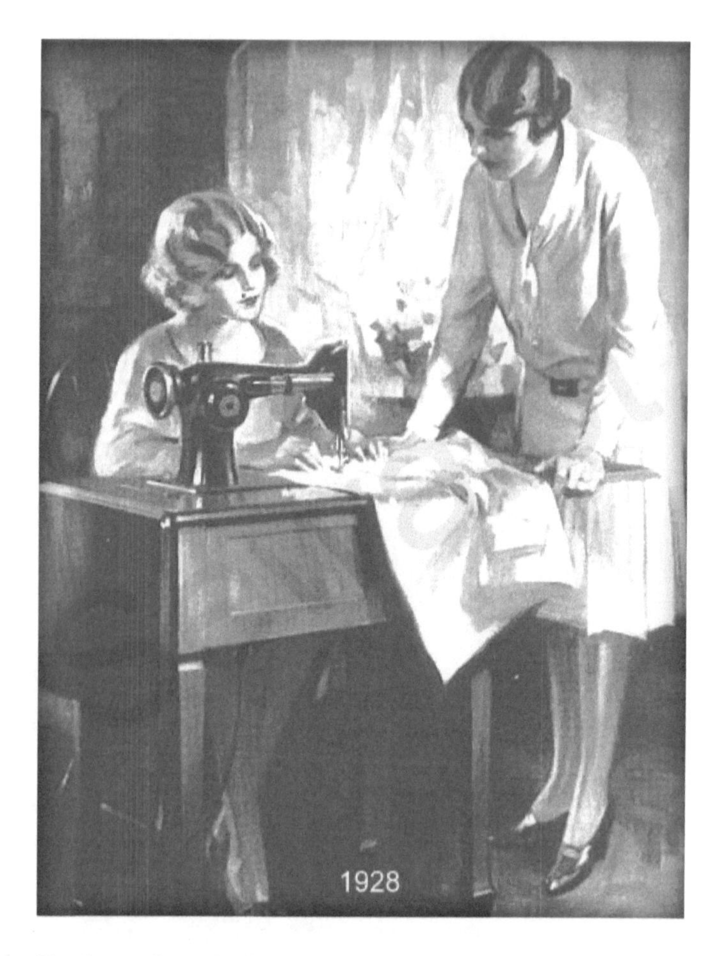

Don't you love the beauty spot! Where have they all gone? This may be an early 201 prototype, possibly a 101 with internal power drive to the integral motor.

From 1935 the Singer 201 soon earned a reputation for reliability and stitch quality that was second to none. Many households dressed their children and fed their families from money earned on this fabulous model. However all this came at a price as we are going to find out.

One of the greatest sewing machine factories on the planet in Clydebank, Scotland. Along with America they produced many of the Singer 201 machines.

Here I am with Singers other special domestic, the amazing Featherweight 222k.

By 1939 the Singer 201 was hitting the streets in big numbers. People loved the 'professional' machine. It was cast iron, heavy and superb with the added option of an electric motor for an extra £5.17s. Or you could buy it complete in its cabinet as shown here, already electrified. The modern world was here. Although the majority of houses in rural areas still did not have electricity it was sweeping through the streets, towns and cities. An electric machine was three times faster and, as with a treadle, you could use both hands for sewing!

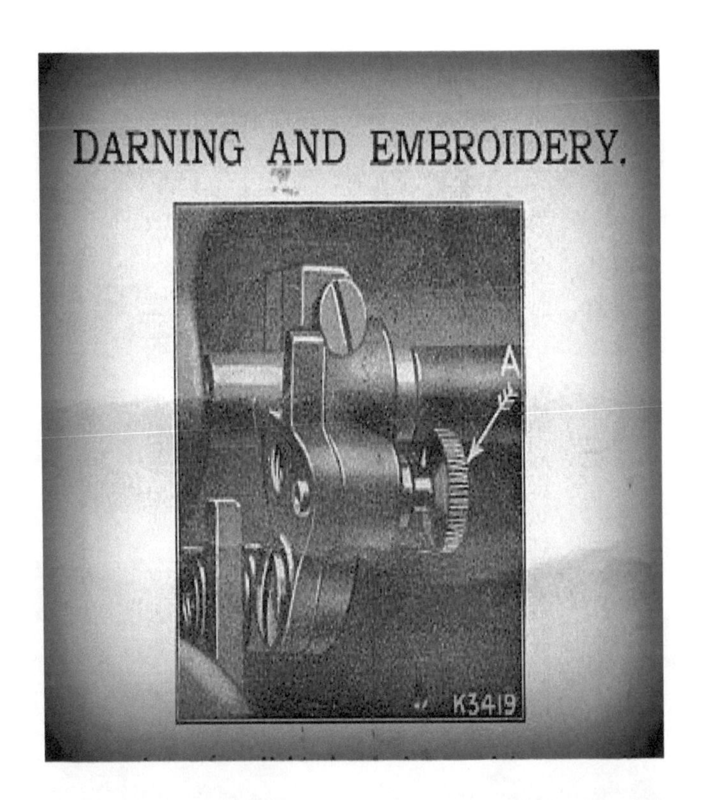

The Singer 201 had a drop feed thumbnut under the bed for free motion embroidery and darning. If you lay your machine back you will spot the thumbnut 'A'. It may be seized but a little releasing fluid and effort will soon have the feed dropping. The large bed and huge clearance made the 201 the perfect darning-free motion embroidery machine.

The Singer model 201 first hit the streets during the depressed 1930s. The worldwide slump was a difficult time for many. It is interesting to think that Singers came out with their most expensive domestic model during the worst possible time.

Singer's famous gears. They seem to have an endless lifespan. This set is over 80 years old and still look new! Interestingly the Singer Featherweights use similar gears to transmit the power through the machine as well.

Besides its outstanding hook, the Singer 201 used the highest quality corkscrew harden high-carbon steel meshed bevel cut gears. These were pretty much indestructible and had been used for decades by the top sewing machine makers. Bradbury, in the 1880s, used almost identical gears on their No 2

Rotary sewing machine. Singer had used them on their model 12 back in 1865. In a lifetime in the sewing trade I have only ever seen two damaged gears on a 201. They smoothly, quietly and efficiently provide the circular movement from the top shaft all the way to the hook. This perfect transfer of power allowed for high speed sewing with minimum vibration. Something that new machines today can only dream about.

For the first time stitches could be set to SPI, stitches per inch. This was a great benefit in mass manufacturing of garments. Today, British Standards require the same tolerance (if you are lucky enough to be certified by them).

The stitch length was via a lever and slot assembly allowing for the first time exact stitch length, time and time again. If you set your stitch length lever to eight stitches per inch, you got eight stitches per inch. **Fully up on the lever provided reverse feed**.

With a full rotary hook, drop in bobbin for easy threading and a superb feed mechanism the Singer 201 had a wonderful silky smoothness. NOTE: A DROP OF OIL ON THE HOOK AND THROUGH THE CENTRE HOLE EXTENDS LIFE. The Singer 201 hook assembly is a work of art in engineering. It is made to tolerances so fine that even a single thread in the wrong place could jam it solid.

The Singer 201 takes the same bobbins as the Singer 66, 99, 185 and many more models. By the beginning of the 1960s Singer swapped to plastic bobbins. A plastic bobbin in a Singer 201 gives a lighter tension and can work perfectly if adjusted properly.

CHAPTER 5
The World's Most Perfect Sewing Machine, the Singer 101

Singer had been busy designing their new machines for years, slowly perfecting the mechanism. They used their experience gained from the purchase of the Wheeler & Wilson business in 1905 to improve on their amazing rotary hook.

For the convenience of the Public there are

SINGER SHOPS

in every City for the sale of

SINGER and WHEELER & WILSON SEWING MACHINES.

Old Machines of any make exchanged.
Easy Terms of Payments arranged.
Liberal Discount for Cash.

After the Singer Company purchased Wheeler & Wilson they had the use of their fabulous Bridgeport plant and engineers. By 1926 they were ready to launch their luxurious Singer model 101.

By 1926, Singers had nearly 1,000 shops in Great Britain and Ireland alone. Worldwide it was the largest company there had ever been, employing countless thousands of people.

Crazy to think that it all started with a bet between two people back in 1850. The tale of Isaac Singer is an extraordinary one and well worth a read if you ever get the chance.

My book on Isaac was a No1 New Release and No1 Bestseller on Amazon when it came out. The limited edition hardbacks were selling for over a $1,000 at one point. I'm always amazed that his story has never been made into a Hollywood blockbuster. Isaac's story is stranger than fiction, from his birth on the American Frontier to his death at his palace in Devon, England.

The Singer Model 101

The Singer domestic model 101 came before the 201. It had a full rotary hook running horizontally in the bed and took a Singer 66 drop in bobbin.

Advertised by Singers as 'the world's most perfect sewing machine' it was the forerunner of the 201. Much of the inner workings were different to the 201 but there were also many startling similarities. The 101 had an automatic lubrication system from an oil reservoir and a detachable base.

The 101 came in a dozen subclasses, depending on minor differences. The Singer 101 was a workhorse and ran from early 1926 up to the early 1930s. It gave Singer the ideal testing ground for their next ambitious machine.

One of the interesting points about the Singer 101 was that it was Singers (and the world's) first mass-produced 'fully electric' domestic sewing machine with a built in 'direct drive' motor. You had no option but electric with the 101. Speed was controlled by a knee mechanism. Strangely, once again Singer did not put a reverse on the Singer 101 model.

This image is from the 1926 Singer 101 brochure. The Sewing Machine 'De-Luxe' for the modern woman. Note how the motor is at the bottom! I'm pretty sure this was an artist's mistake or 'poetic licence'. Maybe a few with belt drives will turn up.

By 1932 around a quarter of a million Singer 101 sewing machines were produced, many are still running today.

The Singer 101 cabinet doubled as a nice hallway table.

Singer kept developing their rotary hook system and getting it better and better. By the 201 it was perfect and is still widely copied today. Almost every 'drop-in bobbin machine' made use of the principles from this amazing hook.

One interesting characteristic of the hook is that it picks the stitch up from the left of the needle as you are facing it. NOTE: MODELS 201, 1200, 221 & 222k NEEDLE: FLAT-SIDE TO OUTSIDE, THREAD FROM RIGHT TO LEFT. However, due to the price of the Singer 201 hook, no one has ever copied it exactly. Only Singer, with its enormous manufacturing abilities could make it cheap enough to be financially plausible.

The narrow foot (only 9mm wide) and small needle hole, made the 201 a quilters dream. Anyone working with fine fabrics loved the way the 201 held and fed the work without chewing up the fabric. Interestingly the Singer 221 and 222k Featherweights both use the same narrow 9mm foot which machinists adore.

The narrow foot and small hole for the needle to pass through the needle plate was possible because the Singer 201 only did a straight stitch. This meant the needle-bar could be clamped into its position allowing only up and down movement. Modern machines have wobbly needle bars that often follow the grain of the cloth, giving you an erratic drunk looking stitch.

The impressive workload that the 201 could handle in varying thicknesses was also made possible by the hardened, forward-facing feed dogs. The indestructible steel used for the feed dogs was apparently a secret formula of Boron and Manganese mixed in an Argon gas crucible furnace.

Due to their shape, cut and design, the teeth could move almost any normal work through the machine.

Today most makers use soft pointed teeth that wear easily and can never be relied on for exact stitches. This is in part due to a modern machines ability to do multi patterns. New machines can do many stitches. What the 201 machine does is one stitch, perfectly.

The needle plate, like all the bright work on the 201 was chromium plated. A method I believe that Singer had perfected around 1929 in their Kilbowie Plant. All Singer 201 models were chrome plated. Chromium plating was more durable that nickel plating and far harder wearing.

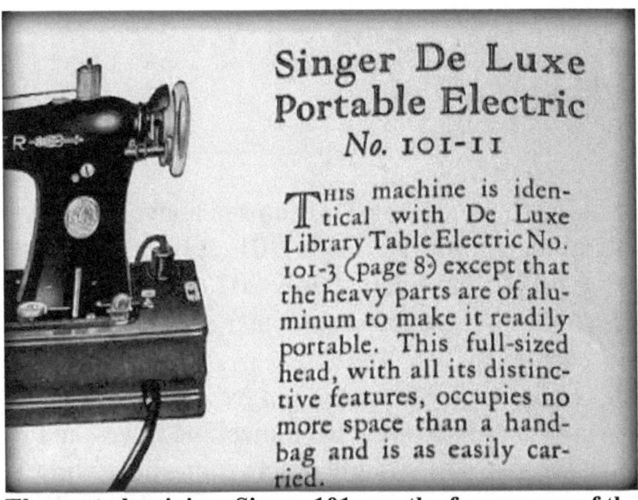

The part aluminium Singer 101 was the forerunner of the 201, different but similar. It had no reverse, a stitch dial and a potted motor. It came as a portable, with a knee speed control, or in a stand with a knee presser bar lift and pedal speed control.

The Jubilee Sewing Machine

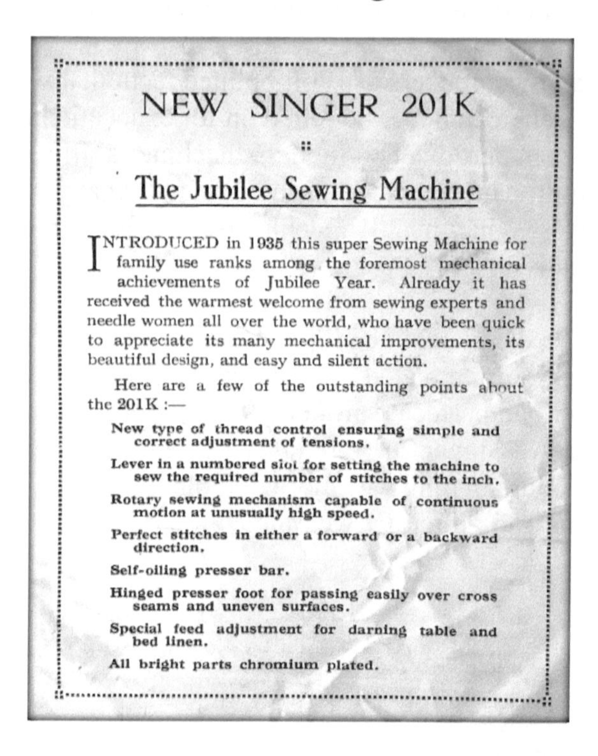

NEW SINGER 201K

::

The Jubilee Sewing Machine

INTRODUCED in 1935 this super Sewing Machine for family use ranks among the foremost mechanical achievements of Jubilee Year. Already it has received the warmest welcome from sewing experts and needle women all over the world, who have been quick to appreciate its many mechanical improvements, its beautiful design, and easy and silent action.

Here are a few of the outstanding points about the 201K :—

New type of thread control ensuring simple and correct adjustment of tensions.

Lever in a numbered slot for setting the machine to sew the required number of stitches to the inch.

Rotary sewing mechanism capable of continuous motion at unusually high speed.

Perfect stitches in either a forward or a backward direction.

Self-oiling presser bar.

Hinged presser foot for passing easily over cross seams and uneven surfaces.

Special feed adjustment for darning table and bed linen.

All bright parts chromium plated.

As I mentioned earlier, you can see above, the official launch for the Singer 201 'Jubilee' was in the year 1935 but interestingly 201's were made before 1935. More about that later.

Sales really took off for the Singer 201 just before World War II and boy what a machine it was, quite possibly the finest straight stitch domestic sewing machine in the world at the time (arguably it still is).

However by 1939 Singer was busy with the war effort. All but essential sewing machines would have to take a back burner.

CHAPTER 6
Singer, The War Years 1939-45

During WW2, the Singer factories around the world, like most manufacturing companies, helped with the war effort, depending on which country the factories were in.

I've been informed that the factory in Wittenberge, Germany, was turned over to the Nazi war effort. After the war the factory found itself inside the Russian zone of Eastern Germany. Much of the factory was stripped and taken back to Russia for war reparations. Interestingly the machinery was later taken to the Russian Singer plant at Podolsk.

The Singer 201 during the war was a rare machine indeed as few were made. During this period serial numbers become erratic and strange dates start to appear as old stock was used up. Picture courtesy of Mrs Milligan, one of my customers.

Out of all the 20 or so Singer factories around the world the two biggest and most important were the Elizabethport plant in New Jersey and the Kilbowie site at Clydebank, Scotland, closely followed later by the Podolsk plant in Russia. The two main monsters in America and Scotland often supplied smaller plants with parts to help them complete their machines.

Now, in WW2 really interesting things happened at Kilbowie and Elizabethport. But before we go to Scotland let's look at the American plant and what they were up to.

The Holy Grail of Handguns

On the eve of the Second World War in 1939, the United States War Department gave a special contract (Order W-ORD-396) to the Singer Manufacturing Company. It was to build a limited number of handguns. It was a test really, to see if the Singer Elizabethport, New Jersey factory could build a gun of suitable quality and numbers. The Singer Sewing Machine Company had built .45 automatic pistols during the Great War, along with 75mm Cannon. The company proceeded to tool up and build 500 .45 calibre Browning designed pistols, model M1911A1, serial numbers S800001-S800500.

Singer were super high quality engineers and worked to tolerances that few other companies did. The result was that, although Singer initially found it hard to produce 100 pistols a day on their production run (the number the War Department had asked for) they did produced 500 of the finest handguns ever made. On examining the pistols the War Department quickly realised that the Singer Company could manufacture far more important 'high precision' items for the war effort. By 1940 Singer were retooling and making specialised aircraft components.

Amongst the many items that they produced were targeting and navigational equipment for the Sperry T-1 Bomb Sight. They made gunfire control computers for the Boeing long range heavy bomber, the B-29 Superfortress. They also manufactured an assortment of precision parts, from directional gyros for planes, to automatic pilot systems. Even artificial horizon instruments for the cockpit. The most important by far were the navigational components that allowed the aircraft crew to find their way safely home.

By 1942 The Singer Company in Elizabethport, had almost stopped making sewing machines as they concentrated on the war effort. They resumed in 1945 with 20,000 basic straight-stitch domestic sewing machines (models 15 and 66). It is said that by 1945 Singer had back orders for over 3,000,000 sewing machines!

But what happened to their superb original Singer pistols from the Elizabethport factory? The 500 pistols were distributed amongst the Army Air Force aircrew. The few that survive today have become legend. They are one of the most iconic handguns ever made and the Holy Grail for gun collectors.

At the beginning of December 2017, The Rock Island Auction Company of Illinois, America, sold a Singer pistol for an astounding $414,000. It is one of the highest prices ever paid at auction for a handgun.

Now let's look at Kilbowie

Now something fascinating happened at Kilbowie during this period. The plant was under American control from New York Head Office. Of course in the early part of WW2, America was neutral, only jumping in after the Japanese attack on Pearl Harbor in December of 1941.

To maintain neutrality, the Singer Company could not make parts for the fighting nations. To get around the parent companies refusal to carry out munitions work, the Kilbowie factory was nationalised for the duration of WW2 by the British Government. This cleared the way for the giant factory to produce vital munitions.

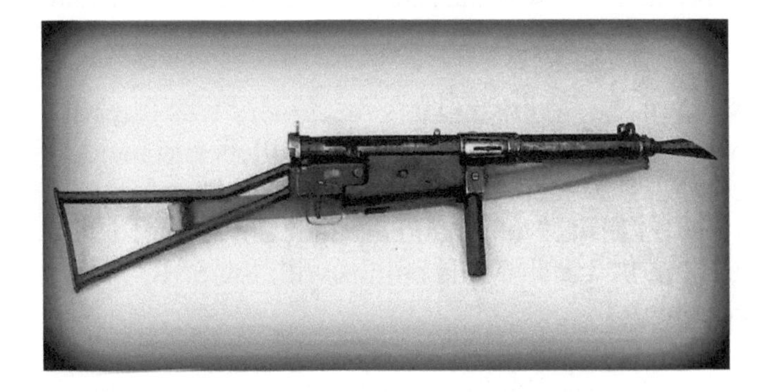

A super rare Mk1 sten gun with kind permission from Robin Storey

By 1941 the Kilbowie plant was making a Singer MK1 sten gun. They made approximately 300,000 MK1 and MK2 sten guns at the Scottish plant. They were issued mainly to the Home Guard but a few were dropped to the French Resistance and a few more found their way to the Western Desert campaign.

Mk1 sten guns with the Home Guard in 1941. With kind permission from Robin Storey.

Most of the Scottish plants machinery was turned over to production of sten guns, bullets and hand grenades. Singers received their first order for 100,000 Mills No36 hand grenades in July of 1940 and marked their Mills bombs with the SMC company logo.

My wife's grandfather, Cyril Cottington, who spent his time in the Royal Air Force during the war, told me that Singer made a particularly good tangent rear sight for the Lee Enfield service rifle. Grampy was a superb shot, winning a silver medal at Bisley. He told me these sights were always called Singer Sights and dearly sought after by crack shots.

One final point before we put this fascinating subject to bed. Although Singer made around 300,000 sten guns it is believed that less than a dozen may survive. Why? Let me tell you.

Post War Singer Sewing Machines

Britain, like most of war-torn Europe, was trying to rebuild after the devastation of World War Two. However materials were in scarce supply and huge demand. The British Board of Trade implemented strict regulations to allow material to be used in manufacturing. Singer would have to achieve a 50% export rate to allow it to meet its material allowances. If it fell below that, steel and other materials would have been diverted to other more profitable companies.

Once the parent company regained control of its Scottish plant they looked for all and any scrap so that they could meet their export targets. With such close contacts with the War Office it was not long before arms, used during the war years, were being melted down back into sewing machines. Millions of weapons during the next few years were recycled back into raw materials for sewing machines (including the sten guns).

Now we are going to go a bit off topic to Princess Elizabeth and then back to something that most people have never heard of but was very likely part of the first post war Singer 201s. Duralumin.

Princess Elizabeth 1945

The Singer 201 sewing machine. Fit for a Princess and used by one!

Princess Elizabeth and the Singer 201K

In 1947 the Singer Company proudly presented HRH Princess Elizabeth their finest sewing machine, the Singer 201. If you watch the Pathe newsreel of the presents for the Royal Wedding you will see the British 201 Singer proudly on display, next to the jewels and gifts that the couple were showered with, from royalty around the globe.
It was a brilliant piece of marketing from Singers. For a long time people were asking for the same

machine that Princess Elizabeth used. If it was fit for a Princess it was certainly good enough for them! Incidentally that year the 201k-3 cost twenty eight pounds eleven shillings and six pence. A bag of chips at the time cost half an old penny. For the cost of the sewing machine you could have bought 13,676 bags of chips (fries). Chips every day for 37 years! Now that would need some salt and vinegar.

A Singer 201 cast iron sewing machine, advertised in 1951. Notice the rear round scroll plate and bolt on motor with light. Within a few years many British Singer 201s were alloy and a lot lighter. Thanks to the SMC for the image. I was told that the glamorous girl is one of Singers shop demonstrators, not a model brought in for the shot. Someone may even know who she is!

CHAPTER 7
Duralumin

Now, this is only trade gossip but well worth the telling. By the late 1940s there was a big benefit to be had by Singers with all its war contacts. From the end of hostilities, surplus of aluminium (used throughout the aircraft and other military industries) was just ideal for Singers, who traditionally (because of cost) worked in cast iron, (except for their Featherweight model 221).

Spitfires were built extensively from a copper-magnesium-aluminium alloy called Duralumin. It was tougher than aluminium. Amazingly, by 1948 a Spitfire was worth just £25 in scrap. Can you believe it! Because of that, most war planes were soon scrapped. Today, out of countless thousands made, there are only about 60 still flying. Of course all this surplus cheap scrap was ideal for any manufacturer who could recycle the metals.

Singer in Britain apparently managed to obtain large enough amounts of war-surplus aluminium and other raw alloys. This allowed the Singer 201 to be made in alloy for the first time.

The factory retooled and formed new castings and by early 1954 they were producing the first Singer 201k sewing machines in lightweight alloy.

Legend tells that more than one black alloy Singer today may be made of recycled Spitfires, Lancasters and Hurricanes. How amazing is that!

Initially in 1954, the alloy 201 came in black, like its pre-war cast iron counterpart, but within a year had switched to a more modern 1950s two-tone brown. **From my research the black alloy Singers were made for a short period in the middle 1950s. This makes them the rarest of the 201s. The latest I have come across was 1958.**

Well, wasn't that interesting? How Singer adapted during WW2 and survived is fascinating to me. Of course not all its plants did but Kilbowie and the Singer 201 were on a roll. Order books were full and sewing machines were once again rolling off all the Singer assembly lines.

The waiting lists for new machines were dropping as Singer filled the demand on the high street. The Singer 201, that everyone thought would have stopped forever had, in the intervening years, become legend.

The full rotary hook and high-carbon super-hard corkscrew meshing gears, all set in a lighter alloy body, gave the machine a smoothness that other sewing machine manufacturers could only dream about. But all this came at a cost. It soon became one of Singers most expensive domestic sewing machines.

The Singer 201 was bought mainly by professional machinists who would sew for a living. The Singer 201 was pretty much bullet proof and could take a direct hit from a nuclear missile and still sew. There is still no domestic sewing machine made that will stitch better than the Singer 201 sewing machine that is in good working order.

Singers 201 was undoubtedly the pinnacle of Singers quality machines. In 1935 the 201 had arrived on the scene after 80 years of evolution. It had the very best features of all Singer models and a price-tag to match. Now would be a good time to start talking about Singer serial numbers.

The Singer Centenary Badge of 1951 proved such a success that sales for machines with this badge rocketed. Singers rushed to fill orders and used up many stock models that were not actually made in 1951. People often find that the manufacturing serial number on their machine does not match their 'special' 1951 badge. These are genuine machines used by Singers to fill the shortfall of orders. By 1952 no new factory machines used the Centenary badge as customers would assume that they were being sold 'old stock'.

CHAPTER 8
British Serial Number from 1930-1966

Because it is now so easy to Google the serial number of your machine today (just type in Singer 201 serial numbers on any computer) I luckily have not had to spend endless pages listing 201 numbers. However, I have listed many of the British serial numbers as they were the most numerous and may prove helpful. I have started just before the 201 and finished in 1966. THESE ARE FOR ALL SINGERS NOT JUST 201.

Please note this is only a guide, not gospel! Some people mail me to say they have a receipt from 1950 so how could my guide have their machine as made in 1948? Let me explain.

The production runs at factories like Kilbowie were complex and long. The castings were marked with the serial number during manufacture. The machines were miles from completion, packing and delivery. Then there is delivery to the depot, storage, sales to the shop, and eventually sales to the customer.

These factors all effect the purchase/receipt date, but not the date of manufacture. For example, during WW2 it is a well-known fact the Singers were making guns and bullets as well as sewing machines. Only when they had spare time would they continue with sewing machine production.

I came across a customer who bought her machine brand new in 1946 yet the casting was clearly made in 1939 just before the outbreak of WW2. During World War Two Singer had back orders for millions of machines! You can see from the numbers how low the production was during WW2. Also it is interesting to note how busy they were in the 1950s which were really the final glory days of the original Singer Empire. By the 1960s machines were flooding in from the Far East and by the 1980s the original Singer Company, started in the 1850s, was in freefall.

1930 Y-7.450.267 to Y-8.375.207
1931 Y-8.375.208 to Y-8.449.942
1932 Y-8.449.943 to Y-8.633.634
1933 Y-8.633.635 to Y-9.162.104
1934 Y-9.162.105 to Y-9.633.846
1935 Y-9.633.847 to Y-9.999.999
1935 EA-000.001 to EA-203.878
1936 EA-203.879 to EA-869.974
1937 EA-869.975 to EA-999.999
1937 EB-000.001 to EB-705.753
1938 EB-705.754 to EB-956.428
1939 EB-956.429 to EB-999.999
1939 EC-000.001 to EC-589.135
1940 EC-589.136 to EC-999.999
1941 ED-000.001 to ED-202.377
1942 ED-202.378 to ED-232.773
1943 ED-232.774 to ED-242.053
1944 ED-242.054 to ED-311.246
1945 ED-311.247 to ED-745.856
1946 ED-745.857 to ED-942.976
1947 ED-942.977 to ED-999.999
1947 EE-000.001 to EE-453.220
1948 EE-453.221 to EE-933.528

1949 EE-933.529 to EE-999.999
1949 EF-000.001 to EF-600.940
1950 EF-600.941 to EF-999.999
1950 EG-000.001 to EG-312.860
1951 EG-312.861 to EG-999.999
1951 EH-000.001 to EH-012.026
1952 EH-012.027 to EH-787.882
1953 EH-787.883 to EH-999.999
1953 EJ-000.001 to EJ-449.138
1954 EJ-449.139 to EJ-999.999
1954 EK-000.001 to EK-123.026
1955 EK-123.027 to EK-992.399
1956 EK-992.400 to EK-999.999
1956 EL-000.001 to EL-999.999
1956 EM-000.001 to EM-015.256
1957 EM-015.257 to EM-999.999
1958 EN-000.001 to EN-970.333
1959 EN-970.334 to EN-999.999
1959 EP-000.001 to EP-771.032
1960 EP-771.033 to EP-999.999
1960 ER-000.001 to ER-999.999
1960 ES-000.001 to ES-238.743
1961 ES-238.744 to ES-999.999
1961 ET-000.001 to ET-179.954
1962 ET-179.955 to ET-999.999
1962 EV-000.001 to EV-019.712
1963 EV-019.713 to EV-602.138
1964 EV-602.139 to EV-999.999
1960 EW-000.001 to EW-005.230
1961 EW-005.231 to EW-020.180
1962 EW-020.181 to EW-024.830
1963 EW-024.831 to EW-030.680
1964 EW-030.681 to EW-038.630
1965 EW-038.631 to EW-045.210
1966 EW-045.211 to EW-054.310

Here are a few of the Singer factory letters around the world. These letter are before the serial numbers.

AA to AT	Elizabeth, New Jersey USA	FA to FY	Beginning 1935
BA to BY	Elizabeth, New Jersey USA	GA to GY	Beginning 1935
CA to CY	Bogota, Columbia	HA to HY	Beginning 1960
DA to DY	Karachi, Pakistan	Ja	1924
EA to EZ	Clydebank, Scotland	JB	1936
FA to FY	Clydebank, Scotland	JC	1948
GA to GY	Clydebank, Scotland	JD	1954
HA to HY	Istanbul, Turkey	JE	1961
JA to JE	St. Johns, Newfoundland Canada	KA to KY	Beginning 1959
KA to KY	Buenos Aires, Argentina	LA to LY	Beginning 1959
LA to LY	Taytay, Philippines	MA to MY	Beginning 1934
MA to MY	Monza, Italy	NA	1951
NA to NC	Anderson, South Carolina USA	NB	1956
		NC	1961
ND to NY	Taichung, Taiwan	ND to NY	Beginning 1963
PA to PY	Karsruhe, Germany	PA to PY	Beginning 1954
		QA to QZ	Beginning 1963
RA to RY	Campinas, Brazil	RA to RY	Beginning 1954
SA to SN	Bonnieres, France	SA to SN	Beginning 1935
SP to SY	Lima, Peru	SP to SY	Beginning 1963
TA to TY	Utsunomiya, Japan	TA to TY	Beginning 1956
VA to VY	Penrith, NSW Australia	VA to VY	Beginning 1957
WA to WY	Santiago, Chile	WA to WY	Beginning 1967
YA to YY	Queretare, Mexico	YA to YY	Beginning 1959

The Singer page above has a piece of faulty information, Wittenberge in Germany is not listed and serial letters JA, JB, JC, JD, JE above should be St John, Quebec, known today as St. Jean sue Richelieu, not Newfoundland.

Singer manufacturing plants and letters

Serial numbers and letters can be very confusing. Note: All the letters in BOLD below are letters that come AFTER the model number denoting place of manufacture.

For example Singer model 201K. K for Kilbowie, serial number ED942.977. Singer 201 made in Kilbowie in 1947.

Example: Singer 211G (model 211 made in Germany). Singer 211U (model 211 made in Japan). Model numbers can be confusing but serial numbers are worse…

Now for some more confusing facts. Singer desperately tried to keep some control of its manufacturing serial numbers. As the plants around the world grew and models flourished it became more and more difficult. By switching the letters around, two identical machines could be identified as to where they were made.

All letters **NOT** in bold lettering below come before the serial number on a sewing machine. Example: **A**123456 - Podolsk. All letters **IN** bold below after the serial number denote place of manufacture. Example 123456**A** - Anderson, USA. Does that make sense? A123456 was made in Podolsk. 123456A was made in America.

A - Podolsk, Russia
A - Anderson, South Carolina, USA
B - Elizabeth, New Jersey USA
B - Bridgeport, Connecticut, USA
C - Wittenberge, (Prussia) Germany
D - Elizabeth, New Jersey USA
E - Podolsk, Russia
E - Elizabethport, New Jersey,
F - Clydebank, Scotland
G - Elizabeth, New Jersey USA
G - Industrial Germany
H - Elizabeth, New Jersey USA
J - Clydebank, Scotland
K - Kilbowie, Scotland
K - Elizabeth, New Jersey USA

K - Industrial, Clydebank, Scotland
L - Elizabeth, New Jersey USA
M - Clydebank, Scotland
N - Elizabeth, New Jersey USA
O - No idea?
P - Clydebank, Scotland
P - Podolsk, Russia
R - Clydebank, Scotland
S - Clydebank, Scotland
S- Podolsk, Russia
SJ - Saint-Jean-Sur-Richelieu, Canada
T - Podolsk, Russia
T - Taiwan
U - Industrial, Japan
V - Clydebank, Scotland
W - Bridgeport, Connecticut USA
W - Wittenberge, Germany
X - Clydebank, Scotland
Y - Clydebank, Scotland

The 201-2 201k-2

There is a possibility that the direct drive, potted motor Singer 201-2 (and the 201k-2) were only cast at Elizabethport in America. Although it was possible to buy a 201k-2 in the UK with 220-240 voltage, sales were very low and may have been imports from the US. The potted motor 201 is a rare machine in Britain.

CHAPTER 9
Prices of the Singer 201

I know I'm always going on about the costs but I find it fascinating. When the Singer 201 first came out in 1935 its basic price in a wooden box was £13.17s (the Singer model 128k was £9.17s 6d, the Singer model 66 £11.10s). The differences do not sound much today but they were huge back then. The price for the 201 was for their hand model. For £5.17s you could upgrade to electric or pay over double for the complete treadle. This price stayed pretty stable from the 1930s right up to the early 1940s when the same basic Singer 201 would vary between £13.7s and £15.8s. Add to this basic price the cabinet of your choice and motor.

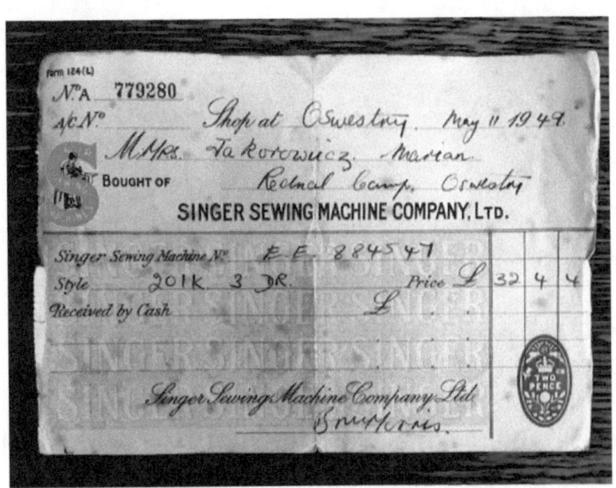

Form 124(L)

No. A 779280

A/c No. Shop at Oswestry. May 11 1949.

M/Mrs. Jakorowicz. Marian

Bought of Redmal Camp. Oswestry

SINGER SEWING MACHINE COMPANY, LTD.

Singer Sewing Machine No. E.E. 884547

Style 201K 3 DR. Price £ 32 4 4

Received by Cash £

Singer Sewing Machine Company Ltd.

B. Morris.

TWO PENCE

Here is a 1949 Singer 201 complete with original receipt. As you can see the machine has rocketed in price to £32, 4s, 4d in an oak treadle. Now the average wage at this point was little more than a few pounds a week in the 1940s and a Spitfire was only £25 in scrap! During WW2 a farm girl was paid 10 shillings a week with board and food. The machine would have represented 64 weeks wages! Most people bought their machines on hire-purchase, layaway or part payment schemes at this time. The 201 Singer represented a huge financial investment, something that has been forgotten down the generations.

1949 Singer 201K3 made in Kilbowie, Clydebank, Scotland.

By the 1950s an electric Singer 201, in its basic wood case (not treadle) was supplied at £20.10s. Whereas pre-war you would stipulate 'electric wanted' and pay the extra, by the 1950s you would have to ask for a hand or treadle machine as the majority were now supplied with motors as standard.

Don't forget the price could almost double if you wanted one of Singers fine cabinets. The Singer factory in Kilbowie had over 2,000 trained cabinet makers.

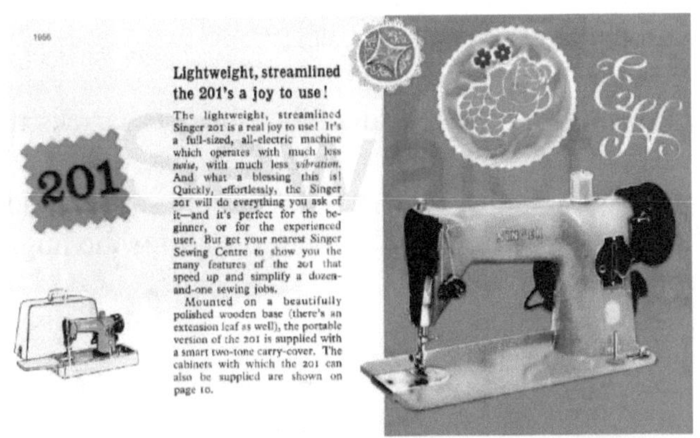

The Singer 201. A joy to use! But what a price!

The 1956 Singer 201K manufactured in Kilbowie, Clydebank, Scotland was £65, 2s, 6d. The massive inflation of the 1950s saw the Singer 201 rocket. By the 1960s they stabilised a little with faster manufacturing techniques. The machine balanced for a while selling in 1961 for £66.

So, in 1956 the Singer 201 was being sold in the UK for over £65. Now, the wage for a well-paid secretary at the same time was £8 a week. The Singer was still costing EIGHT WEEKS WAGES. Today that would translate to around £2,400 for a Singer 201. This extraordinary price was about to come crashing down as the huge manufacturing giants of the Far East came on line (much with American investment after WW2).

Unfortunately this high price from the 1950s onward was disastrous for the Singer Company.

New Far Eastern competitors were starting to supply machines that retailed at a little over a tenner (£10). Of course they were not the same quality as a Singer but we all buy on price don't we. Of course Singers tried offering cash discounts, free lessons, extra feet and other incentives.

Unfortunately, even with all their business tricks the 201 (and the Singer Companies days) were numbered. As the 1960s rolled into the strike ridden 1970s the writing was also on the wall for the huge factory at Kilbowie.

From the 1970s, Singer shops, which once dominated high streets in every major town and city across the world, started to close.

The huge Singer clock at Kilbowie could be seen from almost every house in the area. There was little excuse for the 12,000 workers be late when they lived next to the largest clock in Europe.

The Singer 201 simply represents the very best that Singers could make and it was built to last a lifetime. Even today many professional machinists seek out this elusive model as no new machine can come close to this beauty for reliability and stitch quality.

NEW CHAPTER 10
Rolls Royce and the Singer 201

Another legend I hear often is that Rolls Royce used modified Singer 201 sewing machines to sew its premium super-soft hide for some of their finer car upholstery, dashboards and fitting. The Singer 201 made a small neat hole compared to the larger machines and the feed did not mark the leather. The 201 could never handle seat hide but for the softer-thinner leather parts it was near perfect.

Is this true? I certainly knew several local businesses that switched to the Singer 201 sewing machines from industrials at their factories. I used to service the machines for them and remember clearly the rows of Singer 201s, sewing away so quietly compared to the normal factory din. What I would have given to have a camera with me on my younger days working in the factories.

All the early Singers were cast iron but in 1954 Singer at Kilbowie in Scotland also produced an Aluminium version to reduce weight. Firstly it was black then later in brown.

Here I am sewing a new zip into a leather handbag. The alloy 201 went through six layers of soft leather with a size 18/110 needle. It was hard work but I managed it. You could not sew anything thicker with reliable results. There are few domestic machines that could handle that sort of work without damage.

Remember, whatever adverts tell you there are a million types of leather. Soft thick leather and hard thin leather. The Romans made parts of their shields out of dried leather! The 201 will NOT sew hard or heavy leather without difficulty. Handbag and soft jacket leather is possible.

The first Singer 201 cast iron machines weighed a ton and were known in the trade as 'Back-Breakers'. The alloy 201s were lighter and lost a bit of the ooompf (how do you spell that!) on the heavier work!

The ultimate Singer showroom with the fabulous Singer 222 and the Singer 201, their two flagship machines in 1957.

I remember a joke going around when I was a kid about how difficult it was to get new sewing machines in the 1950s. A woman walks into a Singer shop in post war Britain. "I would like to buy a new Singer." The Singer man just shrugs, "Wouldn't we all madam, wouldn't we all."

In reality times were so hard and material so scarce that they would be put on a waiting list and be notified when one was ready. Some people would wait up to 18 months for a new machine. A little bit different to today eh!

Here I go again…Now, when I say expensive let me explain...

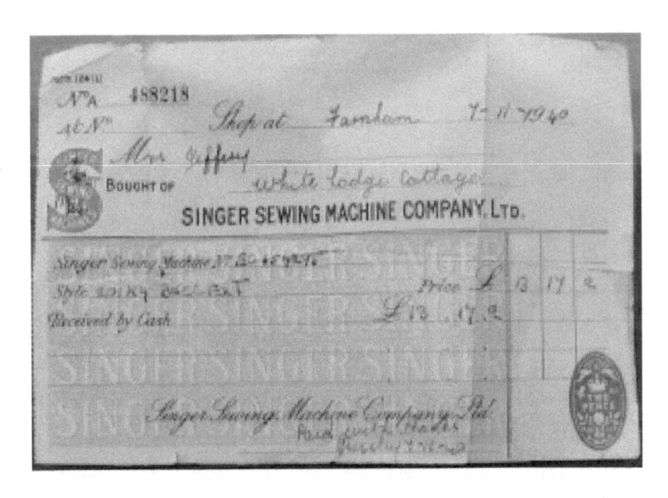

Nº A 488218

Shop at Farnham 7-11-1940

Mrs Jeffery

BOUGHT OF

White lodge cottage

SINGER SEWING MACHINE COMPANY LTD.

Singer Sewing Machine Nº

Style 201K4

Price £ 13 17 2

Received by Cash £13.17.2

Singer Sewing Machine Company Ltd

Singer 201k receipt 1940 when a land girl's wages were 10 shillings a week.

You can see from the receipt that Mrs Jeffrey paid 13 pounds 17 shillings and tuppence for her brand new Singer 201 in 1940. Her weekly wage as a domestic at that time was just over 10 shillings. The average normal wage was a little over one pound. She had paid the equivalent of **27 weeks wages for her** Singer 201K. WOW. **Half a year's wages for a Singer 201 sewing machine!**

Now, in today's money what would that be? Work out what half your years wages are and you have what the Singer 201 cost in 1940. Basically they cost the same as a decent car today! Now you know why the Singer 201K machines sew better than their modern plastic counterparts. I once met a man who paid £50 for his Singer when new. He only remembered because he paid £500 for his house!

The later Singer 201s were angular alloy and varied colours. They were still superb machines but the earlier all cast iron models handled thicker material slightly better because of the clearance and often heavier balance wheel.

The Singer 201 ran for nearly three decades before its production costs made it unviable and production at Kilbowie (and elsewhere) ceased. It was the end of an era and the end of possibly the best straight stitch domestic Singer ever made.

The Final Singer 201 was a browny-beige two-tone and by 1963 as the last stock was sold off from the Singer shops, it was no more.

The finest basic straight stitch sewing machine in the world became just too expensive to make.

One of the many reasons that the 201 did a perfect stitch was the deceptively simple looking tension unit. It is only when the tension is exploded that you can see that there were 11 pieces at work. This included the double secured fully adjustable tension spring, super-hard concave polished tension discs and the automatic pin release that travels through the centre of the tension unit. I have a YouTube clip on tensioning your machine, 'Alex Askaroff explains basic sewing machine tension adjustment'. Check it out.

There are six pieces in the main block on the left assembly and another five on the right. Eleven pieces, all finely balanced and fully adjustable to obtain a perfect stitch through just about any normal material.

I've written extensively about the collapse of the original Singer factories in my books. By the 1960s cheaper imports were flooding the market and the old 201 was put to bed. By 1962 you could buy over 100 sewing machines cheaper than the 201 and the 201 only did one stitch!

Today, as I write early in 2022, the Singer 201 machines still go for silly low prices on Ebay. Few people realise how well they were made or how much they used to cost. Some don't even have the 201 badge, so people sell them without having a clue what they are letting go.

If you enjoy your sewing, grab a Singer 201 while you still can. It is a machine that will make you smile with each stitch.

Note the extra spool pin on the bed. This meant that you could wind a spare bobbin as you sewed, just like an industrial machine. No hanging around on this beauty.

"By any standards, a well set Singer 201 is simply unbeatable for stitch quality. It is undoubtedly the finest Singer domestic straight stitch sewing machine ever produced."
Alex Askaroff

The Singer 201 Sewing Machine, simple perfection.

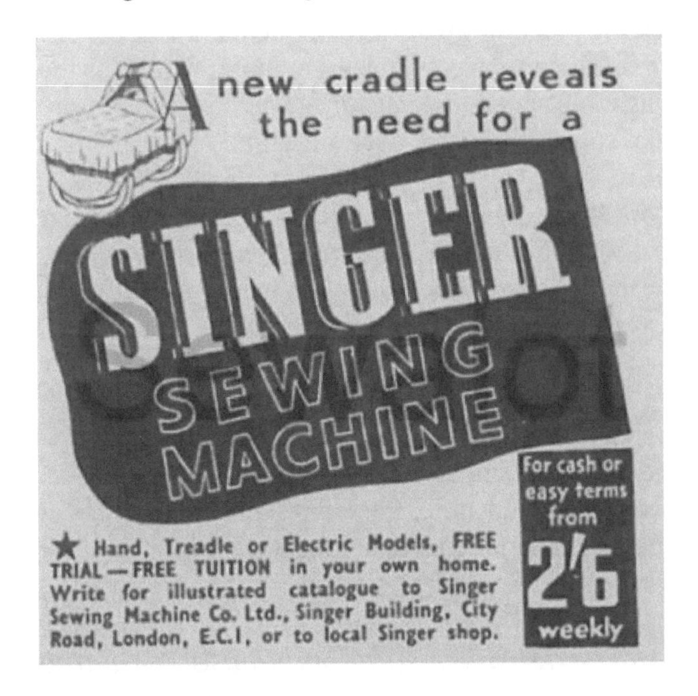

The Singer 201 was so expensive that some people paid over years for their machines. Not like today where some cheap machines will last you a few years before they take up their rightful place on the rubbish mountain. Oh, and in their brief use they will stitch like only plastic can!

CHAPTER 11
The Which Report 1961

In 1961, Which Magazine tested over 40 of the top sewing machines that were on the market at the time. The tests were amazingly in-depth and, as the magazine stated, the machines were expected to take substantial work over a lifetime.

You cannot imagine anyone making anything today that was built to last a lifetime! Where did it all go wrong? For example each electrical foot controller was tested over 75,000 times to review its capability.

Against all opposition and all modern marvels of the time, the Singer 201, won the Which Magazine overall best machine. Remember that Which was an unbiased and independent magazine never weakening to flattery.

After extensive testing and abuse, including dropping the machines... The Singer 201 and Necchi Supernova cannot be seriously faulted.

They were the best but also the dearest machines. The Singer 201 cost £66. 1s. 3d. The Universal 'Sew Maid' machine cost just £11. 19s at the time.

Let’s look at the average wage in 1961. It was well under £20 per week in the UK. So the 201 was still costing around one month’s wages.

Even with all Singers price cuts and streamlining of production it still relates to the sewing machine costing over £1,000 at today's value. Interestingly that is far cheaper than its original 1930s price.

You can see why it was such a beauty! The 201 was costing over six times the price of some of the other sewing machines! It was closely followed in price by the fabulous lightweight Singer 222k at £65. 15s. 6d.

Production of the Singer 201 finished by early 1960s but there was still stock being sold at various Singer outlets as late as 1963 (as the last of the old stock disappeared).

Threading diagram for Singer 201

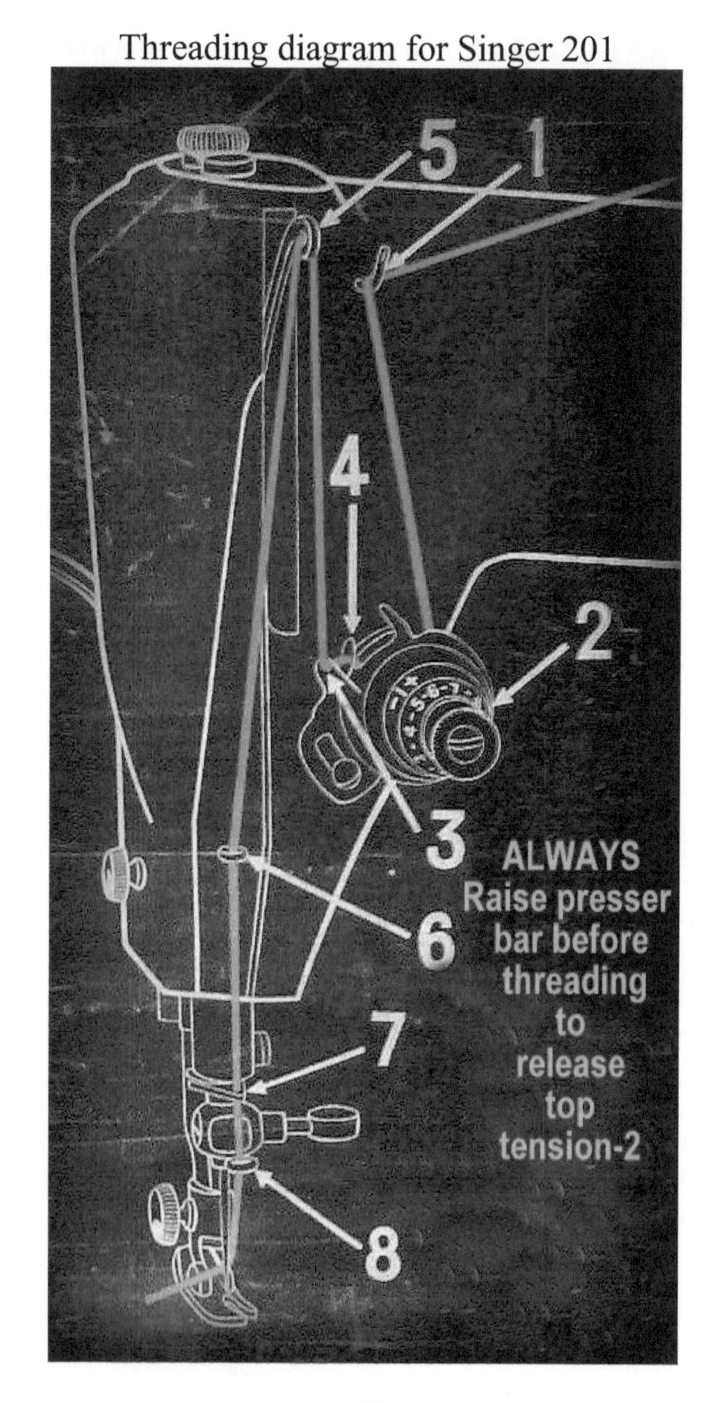

By any standards, a well set Singer 201 is simply unbeatable for stitch quality.
Alex Askaroff

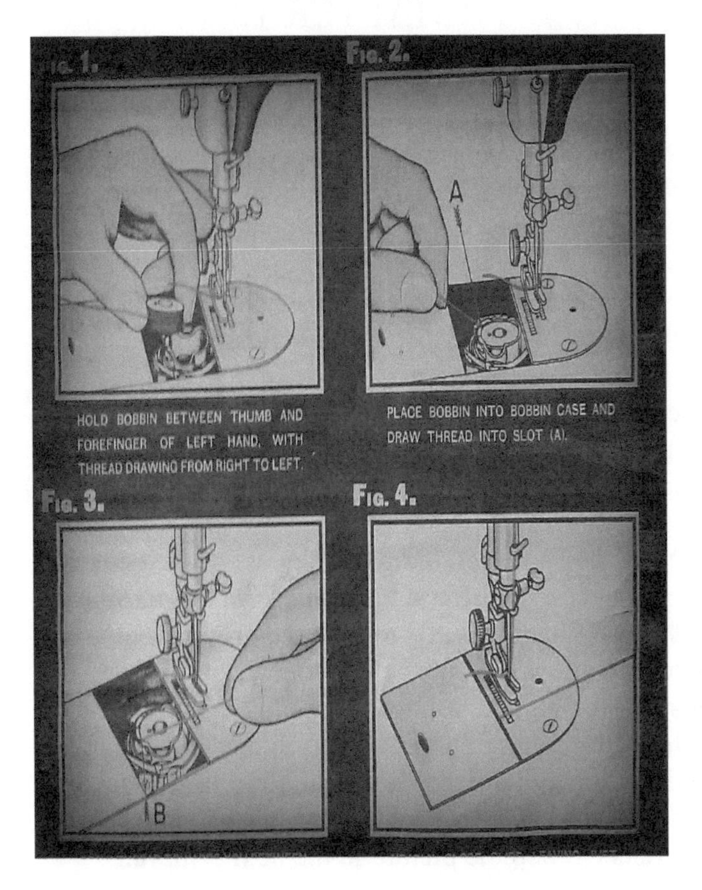

Singer 201 lower threading diagrams

CHAPTER 12
Colours & Cases

Although the Singer 201 changed little during its long production, the machine did wear different clothes.

Not counting dealer modified machines and ones that have had other parts added at some time in their lives, I have come across five distinct changes. For collectors and enthusiast who are going to try and get a 'set'. Here is what I have stumbled over in the last 40 years.

Now, don't forget, today with new decals, your 201 could come in different colours from bright pink to sunrise satsuma. Let's look at the original colours. I tell you now it won't be long before enthusiasts start working their magic on the 201. I've even heard rumours of a mint green Singer 201k out there somewhere, but like a unicorn, I have never seen a real one. I would love to hear of one. If it exists I'll pop the picture in the next print run.

Like the Model T Ford you could have the cast iron Singer 201 in any colour as long as it was black. Cast iron models that ran right from 1935 to around 1962 (depending on who you talk to) and, no matter if it was Britain or America, they were all black.

The cast iron black Singer 201-1 treadle and the 201-4 hand crank ran from 1935 until all production stopped around 1962. It was made in Scotland, Germany, Elizabethport in America (alloy ones were assembled in Australia for a period). It weighed 36lbs without the lid! Pictured is the British hand crank model 201k-4.

The design on the early 201s has become known as 'paper clip'.

The design on the later British alloy singers has become known as prism. However, I was told by a Singer employee that it was the 'infinity' pattern. The workers at the Kilbowie Plant who designed the later pattern (1954-63) would have looked at a machine that their fathers and grandfathers had made at the plant. The 201 seemed to have an infinite life. How true that turned out to be.

The Singer 201-2 or the British 201k-2 with the built in potted direct drive motor on the back and built in light unit on the front. This model ran from 1935 to 1957 in America and 1935-1962 in Britain. It was much rarer in Britain and came with a 240v motor and bulb.

Here is the fabulous alloy Singer 201k-29 hand crank. It was only made at Kilbowie, Clydebank, Scotland and in limited numbers from 1954-1962-3. Only the first ones were black and legend tells that they may contain melted down alloys from old fighter planes. These are the rarest of the 201s to my knowledge and stitch like a dream.

Here is the stunning Singer 201k-3. Only made in Kilbowie after the black alloy models. Unfortunately only the digital version of this book is in colour due to the printing costs. This machine is a lovely coffee and cream two-tone with a deep coffee side cover, stitch plate and balance wheel. The 201k-3 was produced up until 1962-63. See how the same machine can reflect different shades from a different angle.

The very light brown biscuit-beige single colour is one of the most unusual 201s. Only made in Scotland, it came with a bolt on motor and smart accents of chrome on the side-face plate and contrasting coffee stitch plate and even hand attachments. Sometimes the hand or balance wheel was coffee coloured too. Alloy 201s were made between 1954 and 1962 at Kilbowie only. By using your serial number online the year of manufacture will come up. This is one of my favourites.

If you are looking at coffee coloured two-tone Singer 201s online, be careful of the shading. The machines can appear to be very different colours due to natural, fluorescent or LED lighting. If in doubt contact the seller and ask for a natural light image and artificial light image of the same machine. I have ended up with several of the same machines thinking that I was buying unique Kilbowie colours.

Behind is from a batch of 25,000 made in 1955 and the front is from a batch of 15,000 made in 1960. Notice the subtle differences in transfers, badges and shading.

Here you can see the progression from cast iron to mixed alloy to aluminium 1930s to 1960s at the back. All the 201s are amazing performers, the alloy-aluminium ones are 13lbs lighter so they don't break you in half when you lump them around. All could be supplied as hand, treadle or electric right from 1935.

No one is absolutely positive about the total numbers of 201s made from Germany, America, Australia and Britain. It has been claimed that up to a million could still be around. As more information comes to light we should be able to pin down numbers. When the Bonnieres factory closed the paperwork was just dumped.

Here are a few of the bases and cases for the 201. The first bentwood cases made in Kilbowie were said to have been perfected by the craftsmen in the steam room. The Scottish Singer plant employed over 2,000 cabinet makers for their wood and owned their own forests. All the cases were heavily over-engineered and protected the 201s from most damage. These cases are one of the reasons that so many of the machines survive in such fantastic condition.

CHAPTER 13
Values

The values of the Singer 201 have always varied dramatically. New they always cost an arm and a leg but now things are different.

There are a lot of people that the 201 attracts, from enthusiasts and collectors to professional machinists.

Many Singer 201s have no model plates.

One of the big pitfalls (or bonuses depending on your situation) is that many Singer 201s were not marked. No little brass plates were on the machine. Singer usually marked their machines with model numbers, however throughout much of their range some batches came out with nothing on them. Singer never explained why this happened and it was not just the 201 it happened to. From 1851, often Singer models carried no model numbers.

Of course today it is no effort to go online and track down the serial number BUT, for people searching out an elusive 201, people selling them don't always include serial numbers. So if you are looking online and you have no serial number (which can also bring up wrong models) the best way is to know the shape of the 201. By recognising the size and shape of the 201 you are already one step ahead of the pack. SEE PAGE 151.

Values vary dramatically, mainly because of the sellers. Some see their grandmother's prized possession as a solid gold heirloom, others are clearing scrap out of the shed. I have seen Singer 201 machines go for over £800, $1059 in perfect fully working, electrically tested condition. I have also seen them go free to anyone who is willing to pick them up!

Someone advertising an unknown model as 'collection only' is going to get tuppence. Someone advertising a pristine 201 with superb images and all the extras is going to get the top whack.

It is a great time to buy the 201. As I finish this book in early 2022 the machine has only just started to be noticed. People around the world are starting to understand the different 201 models and different styles. They are also starting to realise the quality of the machine, a machine that will last forever and a day.

I added up all the Singer 201 machines for sale in the last week on eBay UK. The average price comes to just over £140, $176. So they are still a steal. Some were as low as £10 collection only (these were all un-badged 201s, so the seller probably never knew what they had). Some were silly high prices (there is always an optimistic seller or two on eBay). No plastic to break or deteriorate, electrics that can be repaired or replaced if needed, or even better a treadle or hand crank. What's not to love!

Today no one thinks twice about repainting machines and new decals are available as well. So you could buy a few scrappers, build complete machines and, after a respray and new transfers, you have a shining beauty worth a mint.

However, collectors come in many groups and those who pay for originality pay a premium for it. So it would be difficult to beat the one-owner cherished machine with original receipts. Look at how some of the early Husqvarna, Bernina and Pfaff machines have rocketed with collectors clubs flourishing, let alone the Singer Featherweight. I used to sell a perfect, fully serviced and tested Featherweight for £100. It would be hard to buy a scrap one for that today.

There is no doubt that certain models will rise above the others. For example all the British made alloy machines are far rarer in America. Then out of the British alloy 201s the light all tan-beige and black are rarer still. Maybe the coffee two tones will rocket! I love their little chrome accents. Very smart indeed sir.

I leave you with one last tip for collecting. There are only three golden rules to 201 values, condition, condition, condition.

Another interesting fact

Molten iron shrinks as it sets. Because the 201 was made to such high tolerances Singer technicians had to work out the exact mix of raw materials for the blast furnaces. Eventually they found the perfect mix that shrunk exactly one inch per hundred as it set from molten to solid. From this they could calculate the exact sizes needed for the perfect machine.

CHAPTER 14
End of an Era

By the early 1960s the Singer 201 had remained virtually unchanged for decades but the world hadn't.

The Far East, mainly with investment from America and other countries, had capitalised heavily in engineering. Unable to make arms or any goods that could be used for war they looked for peacetime household goods to make.

Initially countries like Japan made everything from toys to typewriters. Advances in manufacturing techniques, including automated engineering tools, made the manufacture of sewing machines possible on a huge scale.

Singer found themselves being pounded by not only cheaper imports, but advanced cutting edge (for the time) multi-stitch machines at dramatically cheaper prices.

Singer invested heavily in this new technology but they were still manufacturing in European factories where the cost of living was far higher. Simply, everything they produced cost more to make than their new competitors. It was not only Singers that were affected, all the major European suppliers were being undercut.

The Singer 201, as wonderful as it was, cost an arm and a leg to make. It was never cheap and by the 1960s materials were more expensive than ever.

Singers tried to cut overheads and staff as much as they could. Factories like Kilbowie in Scotland slowly lost more and more staff. Everyone at the plant knew they were in trouble when the huge clock tower was knocked down and sold for scrap!

Singers came out with several straight stitch machines to replace the Singer 201, like the Italian made Singer 239. A simple but efficient machine at half the cost of the 201.

The Singer 239 was made in Monza, Italy and ran on a 15 style bobbin assembly under the machine. It was tough, lumpy and reliable. It was machines like these that would be the replacements for the expensive 201.

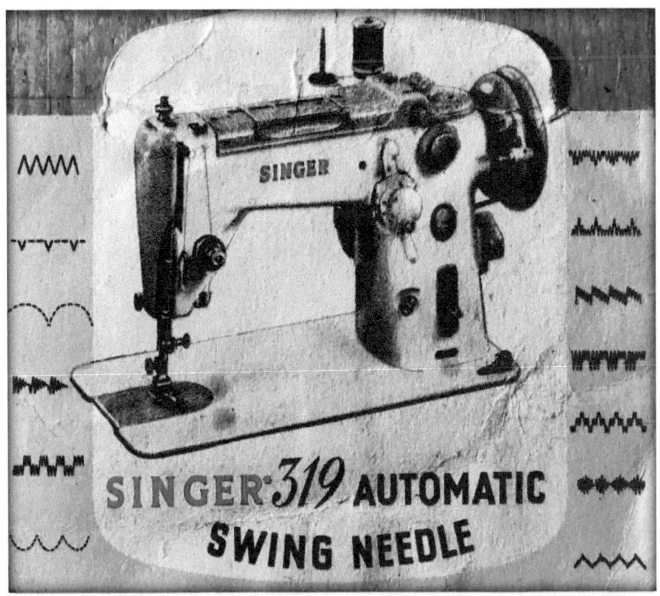

The amazing Singer 319 had also come online, a multi-stitch marvel using several industrial parts. The good old Singer 185k looked at first glances similar to the 201k and far cheaper to produce. In reality the 185 was a Singer 99 in different clothes. A nice sturdy sewer but no comparison to the 201. It would be like comparing Pavarotti to my dad. Yes they could both sing but…

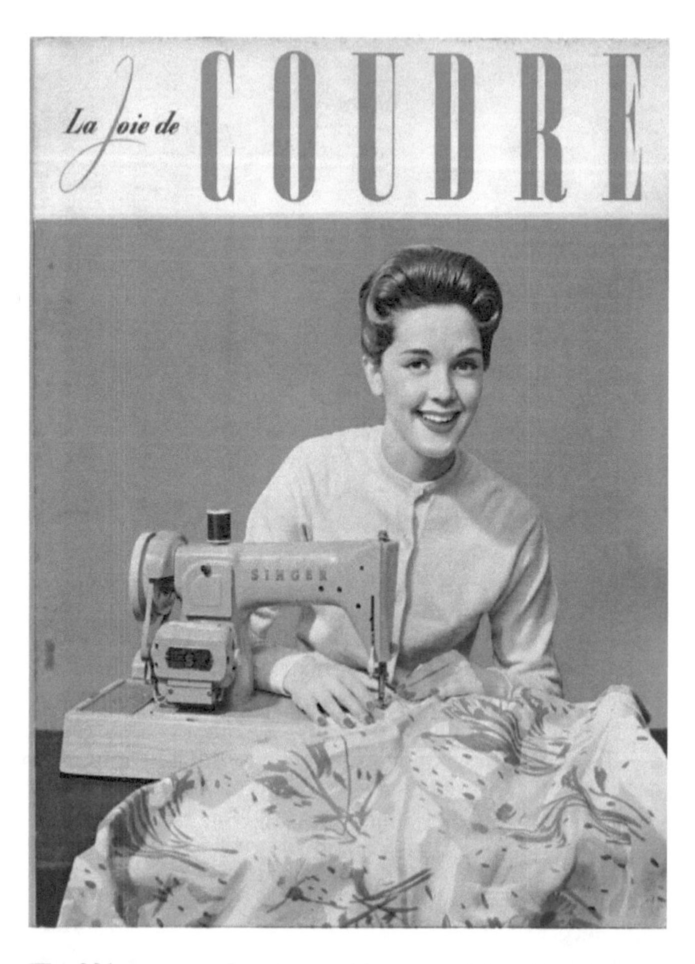

The 201 was popular the world over and the French loved it as much as us. I can't see the front of this model. It might possibly be the French made Singer 191B, made at the Bonnieres Factory not far from Paris. Isn't that the best smile ever!

The 1950s and 1960s saw a rash of new models coming from Singer factories. The 306k from Scotland was a super workhorse incorporating a full industrial rotary hook and it did a zigzag. The German Singer 215G zigzag below came from the Karlsruhe Factory which produced some amazing machines from 1954 until its closure in 1982.

And so with competition coming from all around the world (including Singers own factories) around 1962 the last Singer 201s quietly disappeared from the assembly lines in Britain and the lines were dismantled. It was the end of one of the greatest sewing machines ever made. The 201 faded into history without no song and dance.

Shop windows around the world were filled with all singing and dancing new models in bright colours with a million unknown names.

However like all great legends, the 201 might have been knocked down but it was not out, not by a long shot.

By the turn of the new millennium, machinists around the world, fed up with constantly breaking machines and poor quality stitching, started to look at the old machines. Some of them stood out from the crowd, the Swiss Bernina machines, and the Swedish Husqvarna models (like the wonderful 19e and 21). The German Pfaff machines were there too. There were also dogs! The F&R Homemaker became nicknamed 'the home breaker' because it was so poor. It was abruptly discontinued.

Professional machinists and hobby machinists started seeking out retro beauties. On top of that pile standing proud was the amazing Singer 201. No one had to sell it to the users, it sold itself. It sewed like no other and sung while it did it.

Machinists added new electrics and LED bulbs and smiled while they sewed. Many had forgotten what it was like to get a perfect stitch. They had spent years with plastic machines with huge wide feet and massive slots in the needle plates that ate fine work. They were used to the needle-bars wobbling all over the place creating a roughly straight stitch.

How many swore when they tried to change needles or move bulky work through the tiny spaces under the sewing arm. With the 201 suddenly all that wonderful design, all the years of trial and error, all the ideas that made sewing easier and better were there. A machine created by the finest engineers on earth and made in the greatest factories in the world would become a rediscovered marvel. Not just for now but for all time.

A new generation of youngsters, fed up with Rubbish Mountains and the throwaway society they were surrounded by, sought out quality machines. Quilters fell in love with them all over again. Even collectors started to look at the Singer 201 as an iconic machine that needed to be not only used but displayed. And so the Singer 201 had a new lease of life.

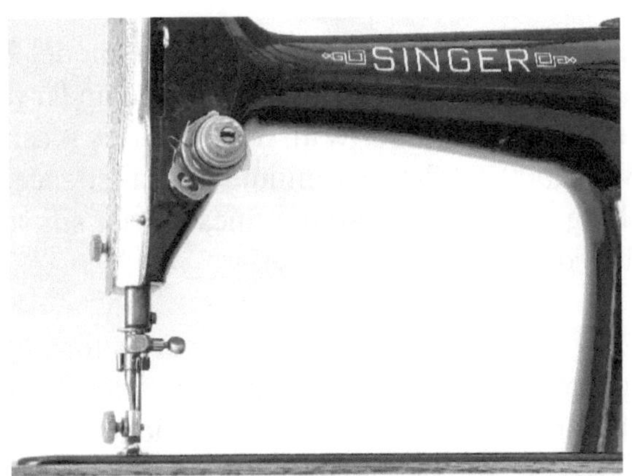

The massive clearance (2cm more than even the Janome Professional under the neck), smooth curves and narrow head assembly makes the 201 a wonderful choice for quilters. Great vision too.

Epilogue

Sew with a Singer 201 in good working order and you will understand what I have been going on about, puuuurfection.

Awhile back I worked with the television production crew for the BBC series of The Great British Sewing Bee. They wanted me to pick an iconic sewing machine to film. Well, in the history of sewing machines there was one model that leapt out to me, the amazing Singer 201. In Series Two, episode Five, all the contestants made their garments using a 201.

My personal favourite was the cast iron model 201 with the wonderful amount of clearance under the bed that quilters adore. With new electrics it can run at over 1,000 stitches per minute in near silence, stitching everything from the finest oyster silk to rustlers denim jeans with ease.

The 201 might not look much, but for its time the design and quality of this machine was on another level (compared to other domestic machines of the period). If it were an Olympic swimmer it would have finished the race before you hit the water!

My dad used to say to me that 'nothing lasts like quality'. The Singer 201, built by families that had worked for generations at the same factories, proved the case.

The Singer model 201 will last forever and a day. In a million years, when only the cockroaches live on Planet Earth, they will be wondering how to use the Singer 201 machines that they stumble over.

Well that's it, not everything there is to know about the Singer 201 but certainly everything I know.

I do hope that you have enjoyed this amazing journey from the Great Depression, to weapons of war, to sewing perfection.

Bye for now.

The ancient worm gear drive, as used on the Singer potted motors, was so successful that even today it is used on heavy vehicles, buses, fire engines, lorries and trucks.

A Perfect Stitch

The Extraordinary Singer 201 'Jubilee' Sewing Machine

By
Alex Askaroff

On The Road Series

There are seven books in Alex Askaroff's **On The Road Series**. They cover his working life around Sussex encompassing a world of stories from the ages.

Book One: Patches of Heaven

Book Two: Skylark Country

Book Three: High Streets & Hedgerows

Book Four: Tales From The Coast

Book Five: Have I Got A Story For You

Book Six: Glory Days

Book Seven: Off The Beaten Track

"If you read any of Alex's 'On The Road Series' you will read them all. They are totally addictive, beautifully crafted and wonderfully inspiring."
Eliza Cooper

Alex Askaroff at Birling Gap

For collectors and enthusiasts
of antique sewing machines and great stories
why not visit

www.sewalot.com

For other publications
By
Alex Askaroff
Visit Amazon

Isaac Singer
The First capitalist
No1 New release

Most of us know the name Singer but few are aware of his amazing life story, his rags to riches journey from a little runaway to one of the richest men of his age. The story of Isaac Merritt Singer will blow your mind, his wives and lovers his castles and palaces, all built on the back of one of the greatest inventions of the 19th century. For the first time the most complete story of a forgotten giant is brought to you by Alex Askaroff.

No1 New Release. No1 Bestseller, Amazon certified.

If this isn't the perfect book it's close to it!
I'm on my third run through already.
Love it, love it, love it.
F. Watson USA

Patches of Heaven is the first book in Alex's popular 'On The Road Series'. We start Alex's working life and follow him as he earns his living. With Nine No1 New Releases on Amazon, Patches of Heaven with enthral readers of all ages.

Alex's second book in his hilarious 'On The Road Series' continues with his travels. We meet forgotten war heroes and crazy customers by the bucket load, from the 1930s debutant balls at Buckingham Palace to a sailor who had a lucky escape from HMS Hood, before its encounter with the formidable Bismarck Battleship.

Once again Alex has worked his magic. These true stories will have you in stitches, you may even shed a tear but you will be left with a happy heart.

Tales from the Coast, Book Four in Alex's On The Road Series, continues the true stories which he brings both England's history and people vividly to life. The stories are as pleasurable as a warm bath after a long day. From the disappearance of Lord Lucan in Uckfield to the Buxted Witch, from William Duke of Normandy to Queen Elizabeth's Eastbourne dressmaker, Tales from the Coast is crammed with a fascinating mix of true stories that will have you entranced from start to finish.

Yet again Alex has woven his magic. I kept saying I never knew that and I'm a local. This may just be one of the best books I've ever read!
J. Vincent

Alex, I've read every book James Herriot ever wrote, and my favorite topics in his books are about the animals and the meals, just like my favorite stories in your books are the ones that talk about your experiences working in people's homes. I love them. Thank you so much.

Joe Edmiston
Louisville, KY

Alex Askaroff has had Nine No1 New Releases on Amazon. For decades Alex has been enthralling readers around the world with his writing. Off The Beaten Track is the seventh and final book in his 'On The Road Series' and completes his working life before retirement.

The 7th Duke of Devonshire's dream started in the middle of the Victorian Era and continues to flourish to this day. World renowned author Alex Askaroff tells the story of Eastbourne in his own unique style, reviving long forgotten characters from the town. We meet local families, fishermen, smugglers, kings and queens, ghosts and even an old witch. It is a tale not to miss.

www.sewalot.com
Alex's antique sewing machine site.

Sir Sewalot, protector of Sewalot.com

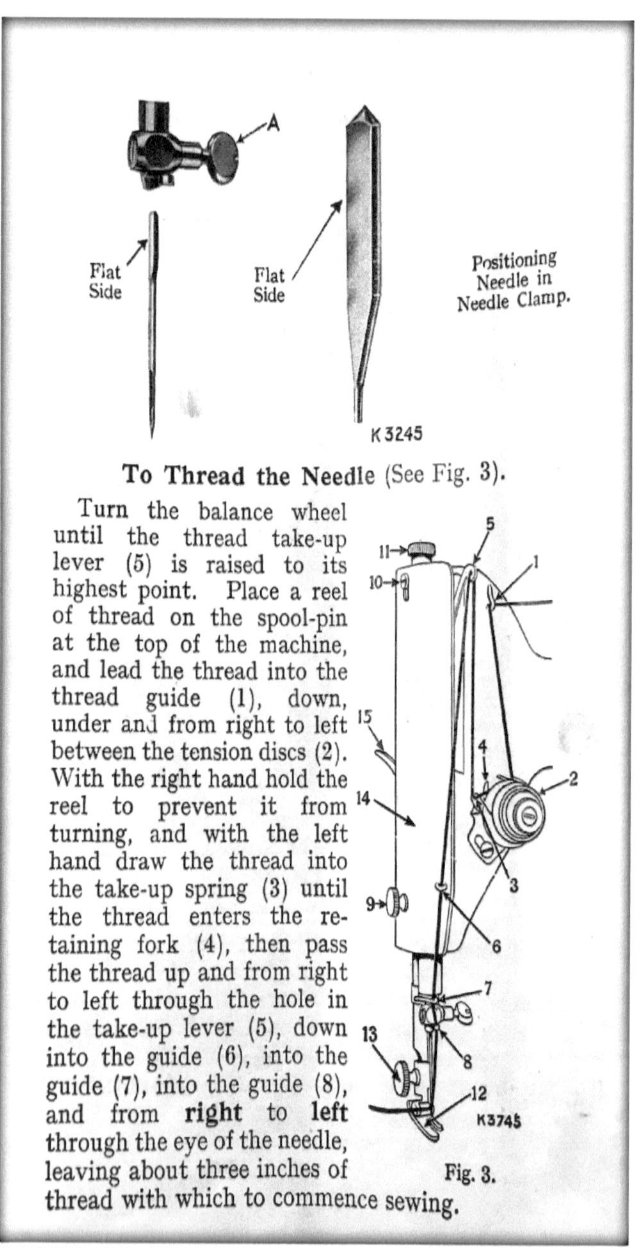

Fig. 3.

To Thread the Needle (See Fig. 3).

Turn the balance wheel until the thread take-up lever (5) is raised to its highest point. Place a reel of thread on the spool-pin at the top of the machine, and lead the thread into the thread guide (1), down, under and from right to left between the tension discs (2). With the right hand hold the reel to prevent it from turning, and with the left hand draw the thread into the take-up spring (3) until the thread enters the retaining fork (4), then pass the thread up and from right to left through the hole in the take-up lever (5), down into the guide (6), into the guide (7), into the guide (8), and from **right** to **left** through the eye of the needle, leaving about three inches of thread with which to commence sewing.

NOTE: FLAT-SIDE OF THE NEEDLE TO THE OUTSIDE. THREAD RIGHT TO LEFT.

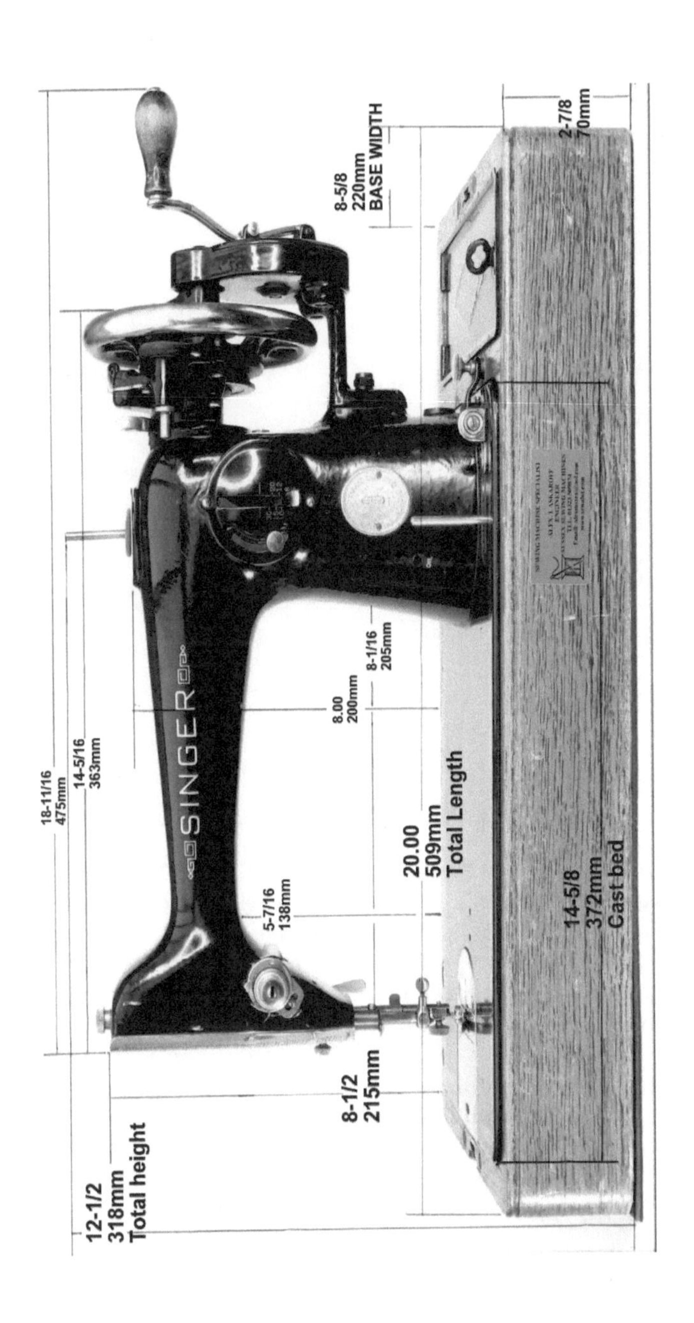

18-11/16
475mm
14-5/16
363mm
SINGER
8-5/8
220mm
BASE WIDTH
2-7/8
70mm
8.00
200mm
8-1/16
205mm
5-7/16
138mm
20.00
509mm
Total Length
14-5/8
372mm
Cast bed
8-1/2
215mm
12-1/2
318mm
Total height

This brown alloy Singer 201 was made in 1957, the year that I was born. In a new plastic case with a modern bolt-on motor and LED lighting it weighs in at 26lbs. The large hollow bases were perfect for the lint and sewing fluff to fall into. With modern machines the sewing debris just jams the mechanisms and gears as there is no space. For a 65yr old it looks pretty good doesn't it! For normal straight sewing, a machine made in the middle of the 20th Century, will out-stitch and out-perform most new machines. With these few modern touches she will be able to sew well into the 21st Century and beyond, creating the perfect stitch. How amazing is that!

The Singer 201 weirdly reminds me of the Aston martin DB5. I know, they are nothing like each other, or are they! I was six when the Aston Martin DB5 was launched in 1963 and it instantly became a classic. A million cars have come and gone and yet in 2021 what car are they using for the James Bond film, No Time To Die? The same DB5 from 1963. Why do we love iconic classics? Why are they timeless? What is the secret of their appeal?

For many of the same reasons, a machine invented back in the early 1930s is still as popular today as it was nearly a century ago. The Singer 201 may not go down in the history of sewing machines as a great machine, in years to come, it may go down as the greatest sewing machine.

Alex Askaroff

Made in United States
Troutdale, OR
08/30/2023